Economics of Organic Farming

Dr. P. Sri Krishna Sudheer awarded his Ph.D. from the School of Economics, Andhra University, Visakhapatnam in 2012. His research interests are concerned Agricultural Development, Production Economics and Rural Development. He is working as Assistant Professor, School of Rural Development, Tata Institute of Social Sciences, Tuljapur, Maharastra. Previously he worked for Andhra University and GITAM University, Visakhapatnam. He associated with International Federation of Organic Agricultural Movement (IFOAM), Germany and Research Institute of Organic Agriculture (FiBL), Switzerland.

Economics of Organic Farming

A Study in Andhra Pradesh

P.S.K. Sudheer

2015

Scholars World

A Division of

Astral International (P) Ltd

New Delhi 110 002

ISBN 9789351308348

Published by : **Scholars World**
A Division of
Astral International Pvt. Ltd.
– ISO 9001:2008 Certified Company –
House No. 96, Gali No. 6,
Block-C, 30ft Road, Tomar Colony, Burari
New Delhi-110 084
E-mail: info@astralint.com
Website: www.astralint.com

Sales Office : 4760-61/23, Ansari Road, Darya Ganj
New Delhi-110 002 Ph. 011-23245578, 23244987

Laser Typesetting : **Rajender Vashist**
Delhi - 110 059

Printed at : **Replika Press Pvt. Ltd.**

PRINTED IN INDIA

The book is dedicated to

Lord Lakshmi Nrusimha,

Antarvedi

From the Office of the IFOAM President
PO Box 800
Mossman, Queensland, 4873
Australia
Email: a.leu@ifoam.org
Mobile Phone: +61428459870
Office Land Line: +61740987610
Skype: andre.leu1

Foreword

It gives great pleasure to write the foreword to "Economics of Organic Farming: A Study in Andhra Pradesh". This comprehensive book was originally written as a Thesis that was submitted by Dr. P. Sri Krishna Sudheer for the award of the Degree of Doctor of Philosophy in Economics to Andhra University, Visakhapatnam, India.

While the book focuses on the economic issues that surround organic farming in Andhra Pradesh, it is set in the context of both international and Indian data.

India has 547,591 certified organic farmers and PGS organic farmers, the highest number for any country in the world. It is rapidly emerging as one of the largest and fastest growing organic sectors in the world. This is includes all facets of the sector especially in terms of increases in the volumes of production and the diversity of crops, increases in the number of producers and the rapidly growing domestic consumer market to drive this growth.This trend is consistent with much of Asia and is being driven by the steady rise of the emerging middle classes who are concerned about food safety, particularly pesticides.

The comprehensive data combined with the original and high quality research that is documented in this book ensures that it will be an invaluable resource document for the organic and sustainable farming sectors in India and globally.

Yours Faithfully,

Andre Leu
President, IFOAM
a.leu@ifoam.org

Acknowledgements

At the outset, I express my deep sense of gratitude to my research director Prof. *L.K. Mohana Rao*, for his extraordinary supervision and valuable suggestions in completion of this research work, without which this work would not have seen the light of the day. I am beholden to him for all that he has done to me all along my stay at Andhra University.

My special thanks are due to Prof. D.S. Prasad, Retd. Professor of Applied Economics, Andhra University for going through the manuscript.

I take this opportunity to thank Prof. S.K.V.S. Raju, Head of the Dept. of Economics and Dean, College Development Council, Andhra University, Prof. K. Sreerama Murthy, Chairperson, Board of Studies (P.G), Dept. of Economics, Prof. R. Sudarsana Rao, Former Head of the Dept., and other esteemed faculty members of the Dept. of Economics, Andhra University, for their encouragement and co-operation in completion of this work.

So many people helped me in completing this work, among them all I should mention M/s. Helga Willer of the Research Institute of Organic Agriculture (FiBL), Switzerland, International Federation of Organic Agriculture Movements (IFOAM), Germany for providing International Statistics and *Centre for Efficiency and Productivity Analysis* (*CEPA*), University of Queensland, Australia for providing me technical assistance. I wholeheartedly thank them for their help.

I am grateful to the Indian Council of Social Science Research (ICSSR), New Delhi for selecting me as a Doctoral Fellow to pursue my Ph.D programme in Andhra University.

I extend my thanks to the Staff of the Agro-Economic Research Centre, Andhra University and the Library of the School of Economics for their help and continuous support during the course my study. I extend also my thanks to the Non–Teaching Staff of the Dept. of Economics.

I am very thankful to my friends Dr. D. Narayana Rao, Lecturer, Govt. Degree College, Narsipatnam and to M/s. R. Vijaya Krishna, P. Rajkumar, A. Vamsi Krishna, D.K. Kumar, N.V.S.S. Narayana, D.V.V. Gopal, CH. Srinivasa Rao, B. Kranthi Kumar, M. Ramesh and other co-scholars for their cooperation and help in the completion of this work.

Finally, I place on record my deep sense of gratitude to my parents Sri. P. Satyanarayana, Smt. P.V.V. Lakshmi maternal uncle and aunt Sri. K. Sridhar, Smt. K.V. Padmavathi and my sister Smt. N. Aparna, my brother-in-law Sri N. Suresh and my brother Sri. P.V.N. Kumar, my sister-in-law Smt. P. Devi for providing me a congenial environment and cooperation from all the directions in completion of this research work.

Last, but not least, I express my deep sense of gratitude to the late Smt. L. Sri Krishna W/o Prof. L.K.Mohana Rao, who showered all her benevolence on me as a foster son and showed keen interest both in my welfare and in the completion of the Work. Had she been alive today, she would have been much elated to see me submitting this Work for my Doctorate!

P.S.K. Sudheer

Contents

List of Tables

Abbreviations

Ac.	-	Acre
AE	-	Allocative Efficiency
AP	-	Andhra Pradesh
APEDA	-	The Agricultural and Processed Food Products Export Development Authority
Bt.	-	Bacillus Thurungensis
CEPA	-	Centre for Economic Efficiency and Productive Analysis
CMA	-	Centre for Management in Agriculture
CRS	-	Constant Returns to Scale
DEAP	-	Data Envelopment Analysis (Computer) Programme
DES	-	Directorate of Economics and Statistics
DRS	-	Decreasing Returns to Scale
EE	-	Economic Efficiency
FYM	-	Farm Yard Manure
FiBL	-	Research Institute of Organic Agriculture
GDP	-	Gross Domestic Product
GMC	-	Genetically Modified Crops
Govt.	-	Government
Ha.	-	Hectare
HYV	-	High Yield Variety Programme
IFAOM	-	International Federation of Organic Agriculture Movement

IIM	-	Indian Institute of Management
IRS	-	Increasing Returns to Scale
Mm	-	Mille Meters
MT	-	Metric Tonnes
NABARD	-	National Bank for Agriculture and Rural Development
NPOF	-	National Project on Organic Farming
NPOP	-	National Programme for Organic Production
SE	-	Scale Efficiency
SFPF	-	Stochastic Frontier Production Function
TCA	-	Total Cropped Area
TE	-	Technical Efficiency
TGA	-	Total Geographical Area
UNDP	-	United Nations Development Programme
VRS	-	Variable Returns to Scale
COAG	-	The FAO Committee on Agriculture
NT	-	Not Tillage
ORG	-	Organic Diversified
NSJV	-	Northern San Joaquin Valley
PCM	-	Pest Control Management
IPM	-	Integrated Pest Management
US	-	United States
EU	-	European Union
IARI	-	Indian Agriculture Research Institute

Chapter 1
Introduction

It is a known fact that Agriculture is the backbone of the Indian Economy. Agriculture in India has a long history, dating back to 10,000 years. Today, India ranks second worldwide in farm output. Agriculture and allied sectors like forestry and logging accounted for 16 per cent of the GDP in 2010, employed 52 per cent of the total workforce and despite a steady decline of its share in the GDP, it is still the largest economic sector and plays a significant role in the overall socio-economic development of India[1]. India faced a severe food shortage when it was unshackled from the clutches of British rule and became independent in 1947. As a result, the Government gave primary importance to Agricultural Sector in the First Five Year Plan. Even then the situation continued till the 1960's. Then the Green Revolution has ushered in in the Country, as a result of efforts of policy makers and agricultural scientists during mid 1960. This Programme aimed at attaining self-sufficiency in terms of food grains, empowering the farmers and modernizing agriculture by using modern techniques and tools to maximize the output of food.

The Green Revolution is one of the greatest triumphs of India. Within a decade, India completely stopped food imports from abroad and no longer was dependent on food aid from abroad. Even if there were food shortages in some parts of the Country, it never resulted in a famine. Thanks to the Green Revolution, India has now emerged as a notable exporter not only of food-grains, but also of several agricultural commodities. Today, India is the world's largest producer of milk, second largest producer of rice, wheat, sugar, fruits and vegetables, and the third largest producer of cotton, just only to mention a few. The direct contribution of the Agricultural Sector to national economy is reflected by its share in total GDP, its foreign exchange earnings, and its role in supplying savings and labor to other sectors. In spite of the advantages accrued to India, in terms of achieving self sufficiency in food production and increasing livelihood choices to the rural poor, Green Revolution made the Indian farmers and those worldover to depend mostly on chemical fertilizers and pesticides, which degraded soil fertility, and environment.

The negative consequences of higher use of chemical fertilisers and pesticides are reduction in crop productivity and deterioration in the quality of natural resources. Pretty and Ball (2001)[2] have pointed out that the environment will be effected by the carbon emission of the agricultural system through: a) Direct use of fossil fuel in farm operations, b) Indirect use of embodied energy for producing agricultural inputs and c) Loss of soil organic matter during cultivation of soils.

Cole *et al.* (1997)[3] have observed that agriculture releases about 10-12 per cent of the total green house gasses emissions which is accounted for about 5.1 to 6.1 Gt CO_2. Joshi (2010)[4] has also pointed out that intensive agriculture and excessive use of external inputs are leading to degradation of soil, water and genetic resources and negatively effecting agricultural production. Arrouays and Pelissier(1994)[5]; Reicosky *et al.(*1995)[6],Sala and Paruelo(1997)[7]; Rasmussen *et al.*(1998)[8]; Tilman (1998)[9]; Smith(1999)[10] and Robert *et al.*(2001)[11], basing on the long term agrarian studies and experiments conducted in EU and North America have concluded that significant quantity of organic matter and soil carbon has been lost due to intensive cultivation

As a result of these changes in the agricultural sector, intellectuals world-over started searching for the ways to come out of the problem of heavy usage of chemical fertilizers and pesticides and finally arrived at to know that organic farming is the only remedy of the problem and also for sustainability of the Agricultural Sector in the long run. In this regard, Kramer *et al.*(2006)[12] pointed out that agriculture has the potential to reduce the emission of green house gasses by crop management agronomic practices. They pointed out that Nitrogen application rates in organic farming are 62-70 per cent lower than conventional agriculture due to recycling of organic crop reduce and use of manure. Some researchers have reported that yields of crops grown under organic farming system are comparable to those under conventional system. Nemecek *et al.* (2005)[13] have also reported that green house gasses emissions from organic farming are 36 per cent lower than conventional system of crop production. In addition, Regonald *et al*(1987)[14] and Siegrist *et al*(1998)[15] have reported that the organic farming system has the potential to improve soil fertility by retaining crop residues and reducing soil erosion. Niggli *et al.*(2009)[16] have reported that the organic farming system has the potential of reducing irrigation water and sequencing CO_2. Mader *et al.* (2002)[17] and Pimental *et al.*(2005)[18] have observed that efficient use of inputs and net income per unit of cropped area on organic farms are at par due to reduction in costs of fertiliser and other input application. Reicosky *et al.* (1995)[19] and Fliessbach and Mader (2000)[20] have pointed out that the organic matter has a stabilizing effect on the soil structure, improves moisture retention capacity and protects soil against erosion. In this context, Pretty and Ball(2001)[21]; Niggly *et al*(2009)[22]have observed that organic farming has the potential to increase the sequestration rate on arable land and in combination with no tillage system of crop production, this can be easily increased by three to six quintal carbon per hectare per year.

As already noted, organic products are grown under a system of agriculture without any use of chemical fertilizers and pesticides with an environmentally and socially responsible approach. This is a method of farming that works at grass-roots level, preserving the reproductive and regenerative capacity of the soil, good plant nutrition, and sound soil management, produces nutritious food, rich in vitality and disease resistant.

1.1 Definition of Organic Farming

An 'organic' label indicates to the consumer that the product has been produced using certain special production methods. In other words, organic is a 'process-claim' rather than a 'product-claim'. An apple produced by practices approved for organic production may very well be identical to that produced under agricultural management practices in vogue normally.

Several countries and a multitude of private certification organizations have defined 'organic agriculture'. In the past, differences in these definitions were significant but the demand for a consistency by multinational traders, has led to great uniformity. The International Federation of Organic Agriculture Movements (IFOAM), a non-governmental organization internationally networking and promoting organic agriculture, has established guidelines that have been widely adopted for organic production and processing.

Most recently, the Codex Committee on Food Labelling has debated on the Guidelines for the Production, Processing, Labelling and Marketing of Organically Produced Foods and adopted a single definition for organic agriculture by the Codex Alimentarius Commission. According to the definition proposed by Codex, "Organic agriculture is a holistic production management system which promotes and enhances agro-ecosystem health, including biodiversity, biological cycles, and soil biological activity. It emphasises on the use of management practices in preference to the use of off-farm inputs, taking into account that regional conditions required,-locally adapted systems. This is accomplished by using, wherever possible, agronomic, biological, and mechanical methods, as opposed to using synthetic materials, to fulfil any specific function within the system."

Organic agriculture is one of the several approaches to sustainable agriculture and many of the standard techniques (e.g. inter-cropping, rotation of crops, double-digging, mulching, integration of crops and livestock) are practised under various agricultural systems. What makes organic agriculture unique, as regulated under various laws and certification programmes, is that: (1) almost all synthetic inputs are prohibited[23] and (2) 'soil building' crop rotations are mandated. The basic rules of organic production[24] are that natural inputs[25] are approved and synthetic inputs are prohibited. But, there are exceptions in both these cases. Certain natural inputs determined by several certification programmes as harmful to human health or the environment are strictly prohibited (e.g. arsenic), while certain synthetic inputs identified as essential and consistent with organic farming philosophy, are allowed (e.g. insect pheromones). A list of specific approved synthetic inputs and prohibited natural inputs is maintained by all certification programmes and such a list is under negotiation in Codex. Many certification programmes also require additional environmental protection measures. Many farmers in the developing world may not use synthetic inputs, this fact alone is not sufficient to classify their operations as organic.

According to United Nations Development Programme (1992)[26] 'Practicing organic agriculture involves managing the agro-ecosystem as an autonomous system, based on the primary production capacity of the soil under local climatic conditions. Agro-ecosystem management implies treating the system, on any scale, as a living organism supporting its own vital potential for biomass and animal production, coupled with biological mechanisms for mineral balancing, soil improvement and pest control. Farmers, their families and rural communities, are an integral part of this agro-ecosystem.

The organic farming in real sense envisages a comprehensive management approach to improve the health of underlying productivity of the soil. Earlier, Lampkin[27] mentioned that organic agriculture is a production system which avoids or largely excludes the use of synthetic compounded fertilizers, pesticides, growth regulators and livestock feed additives. It relies on crop rotation, crop residues, animal manure, legumes, green manure, off farming organic waste and aspects of biological pest control (3).

The most recognised definition of the term "organic" is best thought of as referring not to the type of inputs used, but to the concept of the farm as an organism, in which all the components - the soil minerals, organic matter, micro-organisms, insects, plants, animal and humans - interact to create coherent, self-regulating and stable whole. Reliance on external inputs, whether chemical or organic, is reduced as far as possible. Thus, organic farming is a holistic production system that, takes the local soil fertility as a key to successful production. As a logical consequence, the IFOAM stresses and supports the development of self-supporting systems both on local and regional levels.

1.2 Historical Background of Organic Farming

Although the term 'organic farming' is getting popularity in the recent past, it is surprising to note that it is some 10,000 years old concept! Ancient farmers started cultivation depending only on natural sources. There is a brief mention of several organic inputs in our ancient literature like the Rig-Veda, the great epics of the Ramayana and the Mahabharata and also in the medieval era in Kautilya's Arthasashthra. In fact, organic agriculture has its roots in traditional agricultural practices that evolved in umpteen no. of villages and farming communities over the past millennium. Major milestones in the area of organic farming are presented in Tables 1.1 and 1. 2.

1.3 Status of Organic Farming In The World

Though the roots of organic farming are in India, Organic agriculture is taking rapid strides throughout the World, and statistical information is now available from 154 countries of the world[28]. Its share of agricultural land and farms continues to grow in many countries (Table 1.3). The main results of the latest global survey on certified organic farming[29] shows:

- About Ha. 35 million of agricultural land is managed organically by almost 1.4 million producers.
- The regions with the largest areas of organically managed agricultural land are Oceania (Ha. 12.1 million), Europe (Ha. 8.2 million) and Latin America (Ha. 8.1 million). The countries with the most organic agricultural land are found in Australia, Argentina and China.
- The highest shares of organically managed agricultural land are in the Falkland Islands (36.9 percent), Liechtenstein (29.8 percent) and Austria (15.9 percent).

Table 1.1: Historical Perspective of Organic Farming in India

Ancient period	
Vedic Period (23750 BC)	A mention was made to organic manure in Rig Veda to Green Manure in Atharva Veda. It is stated that to cause healthy growth the plant should be nourished by dung of goat, sheep, cow, water as well as meat. A reference of manure is also made in Vrksayurveda by Surpala (Manuscript, Oxford, No 324 B, Six, 107-164)
Oldest practice	10000 years old, dating back to 'Neolithic age, practiced by ancient civilization like Mesopotamia, and Hwang Ho Basin
The Ramayana (7300 BC)	All dead things - rotten corpse or stinking garbage etc. returned to earth are transformed into wholesome things that nourish life. Such is the alchemy of mother earth – as interpreted by late Sri. C. Rajagopalachari, in his magnum opus – The Ramayana published by Bharatiya Vidya Kendra, Mumbai
The Mahabharata (5500 BC)	A mention was made to Kamadhenu, the celestial cow, and its role on human life and soil fertility.
Kautilya's Arthasashthra (300 BC)	A mentioned was made to several manures like oil cake, excreta of animals etc.
Brihad-Samhita (by Varahmihira) (515 AD)	He described how to choose manures for different crops and the methods best suited for manure.
The Holy Quran (590 AD)	At least one third of what you take out from soils must be returned to it implying recycling or by post-harvest residue

Table 1.2: Key Milestones on Organic Farming in Current Period in the World

Sir Albert Howard (1900-1947)	He is the Father of Modern Organic Agriculture, developed organic composting process (mycorrhizal fungi) at Pusa, (Samastipur) India and published document "An Agriculture Testament".
Rudolph Steiner (1922)	He is a German spiritual Philosopher who built biodynamic farm in Germany.
J.I. Rodel (1950), USA	He popularized the term 'sustainable agriculture' and also method of organic growing.
IFOAM	Establishment of 'International Federation of Organic Agriculture Movement", in 1972
One Straw Revolution	Masanobu Fukoka, an eminent microbiologist in Japan released the Book in 1975.
EU Revolution	EU Revolution on Organic Food, 1991
Codex	Codex Guidelines on Organic Standard, 1999.

Table 1.3: Land under Organic Agriculture in World's Top 10 Countries, (Share of Total Agricultural Land 2005-2009)

		2005		*2006*		*2007*		*2008*		*2009*	
Rank	*Country*	*Organic*	*%*	*Organic*	*%*	*Organic*	*%*	*Organic*	*%*	*Organic*	*%*
1	Australia	11766768	0.026	12345314	0.028	12023135.1	0.028	12023135.1	0.029	12001724	2.88
2	Argentina	2682267.51	0.020	2358375.77	0.018	2777959	0.021	4007026	0.030	4397851	3.31
3	U S A	1640769.203	0.005	1188167.701	0.004	1736084.003	0.005	1948949.128	0.006	1948946	0.60
4	China	2300000	0.004	2300000	0.004	1553000	0.003	1853000	0.003	1853000	0.34
5	Brazil	842000	0.003	880000	0.003	1765793	0.007	1765793	0.007	1765793	0.67
6	Spain	622762.25	0.025	736938.51	0.030	804884.2	0.032	1129843.62	0.045	1330774	5.35
7	India	185937	0.001	432259	0.002	1030311	0.006	1018469.6	0.006	1180000	0.66
8	Italy	1067102	0.084	1148162	0.090	1150253	0.090	1002414	0.079	1106684	8.68
9	Uruguay	759000	0.050	930965	0.061	930965	0.063	930965	0.063	947115	6.26
10	Germany	807406	0.047	825539	0.048	865336	0.051	907786	0.054	930965	5.59
	World	29046687.65	0.007	30144686.28	0.007	32351095.51	0.008	35225259.08	0.008	37232127	0.85

Source: Statistics and Emerging Trends 2010, *The World of Organic Agriculture*, IFAOM, Bonn and FiBL, Frick and also available on world wide web: http://www.organic-world.net/fileadmin/documents_organicworld/yearbook/yearbook-2011/pages-225-sources-world-of-organic-agriculture.pdf

Note: Area in Ha. and % is share in Gross Cropped Area to Total Area Cultivated.

- The countries with the highest number of producers are in India (340'000 producers), Uganda (180'000) and Mexico (130'000). More than one third of organic producers are found in Africa (Table- 1.4).
- On a global level, the organic agricultural land area increased over time in all regions, in total, by almost three million hectares, or nine percent, compared to the data from 2007. Twenty-six percent (or Ha. 1.65 million) more land under organic management was reported for Latin America, mainly due to strong growth in Argentina. In Europe the organic land increased by more than Ha. 0.5 million, in Asia by Ha. 0.4 million.
- About one-third of the World's organically managed agricultural land – Ha. 12 million – is located in developing countries alone. Most of this land is in Latin America, with Asia and Africa in second and third places. The countries with the largest area under organic management are Argentina, China and Brazil.
- About Ha. 31 million are organic-wild collection areas and are land for bee keeping. A majority of this land is in developing countries – in stark contrast to agricultural land, of which two-thirds is in developed countries. Further organic areas include aquaculture areas (Ha.0.43 million), forest (Ha. 0.01 million) and grazed non-agricultural land (0.32 million hectares).
- Almost two-thirds of the agricultural land under organic management is grassland (22 million hectares). The cropped area (arable land and permanent crops) constitutes Ha. 8.2 million, (up 10.4 percent from 2007), which represents a quarter of the organic agricultural land.

The Study of Wyss,H.E *et al*[30] has traced the history of organic farming in Europe and stated that although the European Organic Agriculture Movement was indicated by Rudolf Steiner and Hans Mueller between 1922 and 1940, it was not until the 1980s that pest management researchers began to develop strategies to control pests in organic systems. Today, insect pest management in organic agriculture involves the adoption of scientifically based and ecologically sound strategies as specified by international and national organic production standards. These include a ban on synthetic insecticides and, more recently, on genetically modified organisms (GMOs). The first phase of an insect pest management program for organic systems is the adoption of cultural practices including diverse crop rotation, enhancement of soil quality by incorporation of specific cover crops and/or the addition of soil amendments, and choice of resistant

Table 1.4: World's Top 10 Organic Producers 2007 - 2009

	2007			*2008*			*2009*		
Country	*Producers*	*% in World Producer's*	*Rank*	*Producers*	*% in World Producer's*	*Rank*	*Producers*	*% in World Producer's*	*Rank*
India	1,95,741	14.20	2	3,40,000	24.67	1	677257	37.44	1
Uganda	2,06,803	16.96	1	1,80,746	13.11	2	187893	10.39	2
Mexico	1,28,819	10.56	4	1,28,862	9.35	3	128862	7.12	3
Ethiopia	1,65,560	13.58	3	1,01,899	7.39	4	101578	5.61	4
Tanzania	90,222	7.40	5	85,366	6.19	5	85366	4.72	5
Peru	36,093	2.96	7	46,230	3.35	6	54904	3.03	6
Italy	45,231	3.71	6	44,371	3.22	7	43029	2.38	7
Indonesia	0	0.00	0	31,703	2.30	8	9981	0.55	8
Greece	23,769	1.95	8	24,057	1.75	9	23665	1.31	9
Spain	18,226	1.49	12	21,291	1.54	10	25291	1.40	10
World	12,19,526	100.00		13,78,372	100.00		1809121	100.00	

Source: Statistics and Emerging Trends 2010, *The World of Organic Agriculture*, IFAOM, Bonn and FiBL, Frick and also available on world wide web http://www.organic-world.net/fileadmin/documents_organicworld/yearbook/yearbook-2011/pages-225-sources-world-of-organic-agriculture.pdf

varieties that help to prevent pest outbreaks. In the second phase, habitat management is implemented to encourage populations of pest antagonists. Third and fourth phases of the program include deployment of direct measures such as bio-control agents and approved insecticides. However, the strategies for pest prevention implemented in the first two phases often obviate the need for direct control measures. Approaches to pest management in organic systems differ from those in conventional agriculture conceptually in that indirect or preventative measures form the foundation of the system, while direct or reactive control methods are rare and must comply with organic production standards.

1.4 Status of Organic Farming In India

India is bestowed with lot of potential to produce all varieties of organic products due to its various agro-climatic regions. In several parts of the Country, the inherited tradition of organic farming is an added advantage which resulted in making the Country to stand unonumero in terms of number of organic farm producers and eight in terms of percentage of the of area under organic farming practice to its total area under farming. This holds promise for the organic producers to tap the market which is growing steadily in the domestic market and that related to the export market and cultivated land under certification is around Ha. 2.8 million (2007-08). This includes Ha.1 million under cultivation and the rest is under forest area (wild collection). The Government of India has implemented the National Programme for Organic Production (NPOP). The National Programme involves an accreditation Schemes for certification bodies, norms for organic production, promotion of organic farming etc. The NPOP standards for production and accreditation system have been recognized by the European Commission and Switzerland as equivalent to their countries standards. Similarly, the Dept. of Agriculture (USDA) has recognized NPOP's conformity assessment procedures of accreditation as equivalent to theirs. With these recognitions, all the Indian organic products duly certified by the accredited certification bodies of India are accepted by the importing countries.

India produced around 3,96,997 MT of certified organic products, which include all varieties of food products namely Basmati rice, Cereals, Pulses, Oil Seeds, Tea, Coffee, Spices, Fruits, Herbal medicines, Honey , Processed food and their value added products. The production is not only limited to the edible sector, but also to that of organic cotton fiber, garments, cosmetics, functional food products, body care products, etc. The number of certification organizations in India and their share in the World from 2005 to 2010 are presented in Table 1.5.

India exported as many as 86 items during 2007-08 weighing 37533 MT. The export realization was around $ 100.4 million registering a 30% growth over the previous year. Organic products are mainly exported to EU, USA, Australia, Canada, Japan, Switzerland, South Africa and Middle East. Cotton contributed a major share among the products exported (16,503 MT).

Table 1.5: No. of Certification Organisations in India and World during 2005-10

S.No.	*Year*	*No. of Bodies in the World*	*No. of Bodies in India*	*%*
1.	2005	419	9	2.15
2.	2006	395	10	2.53
3.	2007	468	12	2.56
4.	2008	481	13	2.70
5.	2009	489	16	3.27
6.	2010	532	17	3.20

Source: The Organic Standard and The Agricultural and Processed Food Products Export Development Authority (APEDA) 2010.

1.5 The Problem

As already mentioned, of late, organic farming is gaining momentum in several advanced countries. India is no exception in this regard. Various studies on organic farming indicated that area and products covered under organic farming are increasing at a faster rate in advanced countries while its spread is relatively slow in developing countries like India. It is also evident that the growing demand for organic agricultural commodities in the advanced countries paves way for developing economies for potential export market for organic agricultural products. By international standards, conversion of a conventional farm into an organic farm will take a minimum of three years and during the first two years, the farmer may incur a loss in farming. In this context, a study of economics of organic farming in contrast to the conventional farming may throw light on the problems in the spread of organic farming. It is a fact that India is a developing country and most of the farmers are marginal and small holdings and are operating agriculture at subsistence levels. In this situation, a marginal or small farmer may not prefer to switch over to organic farming from his age-old conventional farming due to the reasons mentioned above. But if he is convinced of the economic benefits of organic farming, he readily accepts to switch over to organic farming. This fact was evident in the case of adoption of HYV seeds in the late 1960's. In turn, such types of studies may also help the policy makers to take appropriate measures to protect the farmer from economic losses in this process of conversion.

1.5 Need For The Study

It is highly gratifying that India achieved self-reliance in food production in the shortest span of time in the World, but despite everything, her traditional agro- system suffered a great setback, especially owing to the indiscriminate use of chemical fertilizers, insecticides, fungicides and herbicides. This has also lead to erosion of soil fertility, contamination of water resources, and chemical contamination of food grain[31]. In addition to this, India has shown interest on the Genetically Modified Crops (GM Crops) like Bacillus Thurungensis (Bt) cotton etc. which are highly hazardous to the environment and also increased her dependence on the foreign seed companies like Monsanto. Of late, many advanced countries like the USA, Switzerland, Australia, Western Europe etc evinced interest in the organic farming practices which generally assure sustainability of agriculture also to the next generation without any compromise on the food needs of the present generation in particular and natural resources like land, water, and environment in general. It is argued that for sustainability of agricultural sector of any country, organic farming is the only way-out as it assures no contamination of water, no environmental pollution and no degradation of soil fertility.

With this back-ground, it can be concluded that there is an urgent need to address this problem in a holistic approach to encourage farmers at the grassroots level to take up organic farming. Also a review of literature revealed that organic farming is beneficial to the human and other living beings by way of providing qualitative food products, protecting environment etc. However, there is inconclusive evidence on the economic gaining/ profitability and economic efficiency of organic farming and there exists a dearth of studies on this aspect in the Indian context. Further, except the pioneering works on organic farming at the CMA[32] IIM, Ahmadabad, which confined their attention to the Northern and Western parts of India, on paddy, wheat, sugarcane and cotton and on the efficiency of inputs used in organic farming and conventional farming and another peripheral study by Prasad[33] which studied several comparative aspects of organic farming and conventional farming, no researcher in India has so far examined location-specific and crop-specific aspects relating to economics of organic farming in a State.

Hence, a comprehensive study dealing with the economics of organic farming and conventional farming covering different agro-climatic conditions is felt necessary. As such, the present Study addressed itself to fill in this gap by examining the Economics of Organic Farming vis-à-vis Conventional Farming in A.P. covering paddy, redgram and groundnut among cereals, pulses and oil-seeds in East Godavari, Mahabubnagar and Anantapur respectively. An attempt has been made in this Study to examine the Economics of Organic Farming in Andhra Pradesh with the following objectives:

1.7 Objectives

The main objectives of this Study are:

1. To examine the trends in the area, production and productivity of the selected crops viz. paddy, redgram and groundnut in the State of Andhra Pradesh and the selected districts of Andhra Pradesh,
2. To analyse the cost of and returns from organic farming practices vis-à-vis conventional farming practices,
3. To assess the economic efficiency of organic farming over conventional farming through the estimation of technical efficiency and allocative efficiency,

   4.To identify the factors determining technical efficiency and
5. To suggest measures that may be useful to the policy makers both at the micro and macro levels.

1.8 Methodology And Sample Design

This Study is based on both primary and secondary data collected from various sources. The sample households for collection of primary data have been selected by using the multi stage stratified random sampling technique. The State of Andhra Pradesh is the study area and three major crops, one each from cereals, pulses and oilseeds viz., paddy, redgram and groundnut have been selected basing on the proportion of area under organic farming. Among the 23 districts of Andhra Pradesh, East Godavari, Mahabubnagar and Anantapur have been selected as they are predominantly cultivating the selected crops under organic farming respectively, which also represent the three natural geographical regions of Andhra Pradesh viz., Coastal Andhra, Telangana and Rayalaseema. In the second stage, 250 paddy cultivating households comprising of 150 organic farmers and 100 conventional farmers households, have been selected from East Godavari District. From Mahabubnagar District, 150 Redgram cultivating households comprising 100 from organic farmers and 50 from conventional farmers households, have been selected From Anantapur District 150 Groundnut cultivating households comprising 100 from organic farmers and 50 from conventional farmer households have been selected. The selection of sampling units in each district for each crop is based on the stratified random sampling technique. The distribution of sample households according to the cultivation practice (Organic and Conventional) has been presented in Table 1.6.

A pre-tested and well designed schedule has been canvassed among the selected sample holdings to elicit information on structure of farm holdings, demographic characteristics, asset structure, cost of cultivation, returns etc. The secondary data have been collected from various issues of Statistical Abstract of Andhra Pradesh and Season and Crop Reports being published annually by the Directorate of Economics and Statistics, Govt. of Andhra Pradesh. The reference year of the Study is 2010-11.

1.9 Techniques Used

Simple statistical tools like averages and percentages have been used in analysing the collected data. Further, Stochastic Frontier Production Function (SFPF) 4.1 and Data Envelopment Analysis (Computer) Programme (DEAP) 2.1 techniques have been employed to assess technical efficiency and allocative efficiency under various situations. In addition, multiple regression analysis has been used to identify the factors determining technical efficiency.

The specification of the above models and the method of estimation of the parameters are presented in the Chapter – VI.

1.10 Concepts Used In The Study

For the purpose of easy grasping and clear understanding, some of the important concepts used in this Study have been presented hereunder:

(a) **Small Farms:** Farms with the size up to Ac. 5.0 have been treated as Small Farms.

(b) **Medium Farms:** Farms with the size from Ac. 5.01 to 10.00 have been treated as Medium Farms.

(c) **Large Farms:** Farms with the size above Ac. 10.01 have been treated as Large Farms.

(d) **Organic Farming Practices:** All those standard farming practices based exclusively on the organic manures, which are locally available natural components like cow dung, neem trees, vermi compost etc. are treated as Organic Farming Practices.

(a) **Conventional Farming Practices:** All those farming practices which apply the chemical fertilizers and pesticides are treated as Conventional Farming Practices.

Table 1.6: Distribution of Sample Households According to Farming Practice and Size of Farm (Village Wise)

Crop	*District/ Mandals*	*East Godavari*									
		S.No.	*Villages*	*Organic*				*Conventional*			
				Small	*Medium*	*Large*	*All Farms*	*Small*	*Medium*	*Large*	*All Farms*
Paddy	**Malikipuram**	1.	Lakkavaram	5	6	3	14	4	3	2	9
		2.	Gondhi	5	6	2	13	3	3	2	8
		3.	Gudapalli	4	5	2	11	3	3	2	8
		4.	Kesavadasupalem	4	5	2	11	3	3	2	8
	Razole	1.	Ponnamanda	5	6	3	14	4	3	2	9
		2.	Kadali	5	6	3	14	3	3	2	8
		3.	Gogannamatam	4	5	2	11	3	3	2	8
		4.	Katranipadu	4	5	2	11	3	3	2	8
	Mamidikuduru	1.	Magatapalli	5	5	3	13	4	3	2	9
		2.	Mamidikuduru	5	6	3	14	3	3	2	8
		3.	Pasarlapudi	5	6	2	13	3	3	2	8
		4.	Pedapatnamlanka	4	5	2	11	3	3	3	9
	Total			**55**	**66**	**29**	**150**	**39**	**36**	**25**	**100**

contd...

Table 1.6 *(contd...)*

Crop	*District/ Mandals*	*S.No.*	*Villages*	*Mahabubnagar*							
				Organic				*Conventional*			
				Small	*Medium*	*Large*	*All Farms*	*Small*	*Medium*	*Large*	*All Farms*
Redgram	Narayanapet	1.	Laxmipoor	5	5	4	14	2	3	2	7
		2.	Ammireddypalle	5	4	4	13	2	3	2	7
		3.	Boinpally	5	4	3	12	2	3	1	6
		4.	Appakpally	4	4	3	11	1	3	1	5
	Kosgi	5.	Kosgi	5	5	4	14	2	4	2	8
		6.	Masaipalle	5	4	4	13	2	3	1	6
		7.	Sampallu	5	4	3	12	2	3	1	6
		8.	Hanmanpally	4	4	3	11	1	3	1	5
	Total			38	34	28	100	14	25	11	50
Anantapur											
Groundnut	B.K. Samudram	1.	Neelampalli	5	6	4	15	2	4	2	8
		2.	Reddipalli	5	6	4	15	2	3	2	7
		3.	Korrapadu	4	4	2	10	2	2	1	5
		4.	Chinnampalli	4	4	2	10	2	2	1	5
	Gooty	1.	Mamillapalli	5	6	4	15	2	4	2	8
		2.	Jakkalacheruvu	4	6	4	14	2	3	2	7
		3.	Thondapadu	4	5	2	11	2	2	1	5
		4.	Ubicherla	4	4	2	10	2	2	1	5
	Total			35	41	24	100	16	22	12	50

1.11 Different Concepts of Cost Of Cultivation:

Cost A_1: Cost A_1 Includes:

i. Value of hired human labour
ii. Value of owned and hired bullock labour
iii. Value of owned and hired machine labour
iv. Value of owned and purchased seed
v. Value of owned and purchased manures
vi. Value of fertilisers and pesticides
vii. Depreciation on farm implements, farm buildings etc.
viii. Irrigation charges
ix. Interest on working capital
x. Land revenue, cess and other taxes paid and
xi. Other miscellaneous expenses.

Cost A_2: Cost A_1 + Rent paid for the leased-in land.

Cost B_1: Cost A_1+ Interest on the value of owned capital assets (excluding land)

Cost B_2: Cost A_1 + Rent paid for the leased-in land + Rental value of the owned land (net of land revenue)

Cost C_1: Cost B_1 + Imputed value of family labour.

Cost C_2: Cost B_2 + Imputed value of family labour

1.12 Concepts of Income

Gross Income: Synonymous with value of output (both main and by product)

Farm Business Income: Gross Income – Cost A_2

Family Labour Income: Gross Income – Cost B_2

Net Income: Gross Income – Cost C_2

Farm Investment Income: Net Income + Rental value of own land + interest on owned fixed capital

1.13 Chapterisation

The present Study has been organised in seven chapters. The First Chapter is an introductory one, which also spelt out the need for the study, the research problem, objectives, methodology used and organization of the work. In the Second Chapter, existing available literature on the studies conducted on the performance of organic farming practice throughout the World and in India is presented. In the Third Chapter, a basic profile of the selected three districts has been presented. In the Fourth Chapter, socio-economic characteristic features of the sample households have been presented. The costs and returns of organic farming practices in contrast with the conventional farming practices have been analyzed by using various standard concepts of costs and returns in the Chapter Five. The Sixth Chapter is the core to the Study, which presents the economic efficiency of organic farming practices vis-à-vis conventional farming practices. Chapter Seven summaries the conclusions of the Study and provides some policy implications for the Study.

Notes

1. Economic Survey 2011, *Planning Commission*, Government of India and for a detailed discussion on the general economic development of India in the recent past, see for instance, Mohana Rao. L.K, budget Meet 2011 held at Dept. of Economics, Andhra University on 5th April 2011.
2. Pretty, Jules and Ball Andrew (2001), Agricultural Influences on Carbon Emissions and Sequestration: A Review of Evidence and the emerging Trading Options, *Occasional Paper, Centre for Environment and Society and Department of Biological Sciences*, University of Essex, U.K.
3. Cole, C.V.; J. Duxbury, J. Freney, O. Heinemeyer, K. Minami, A. Mosier, K. Paustin, N. Rosenberg; N. Sampson, D. Sauerbeck and Q. Zaho (1997), "Global Estimates of Potential Mitigation of Greenhouse Gas Emissions by Griculture," *Nut Cycl Agroecosyst,* Vol. 49, pp. 221-228.
4. Joshi. P.K., (2010) "Conservation Agriculture: An Overview", *Indian Journal of Agricultural Economics,* Vol.66, No.1 pp.53-63.
5. Arrouays, D. and P.Pelissier (1994), "Changes in Carbon Storage in Temperate Humic Soils After Forest Clearing and Continuous Corn Cropping in France", *Plant Soil,* Vol.160, pp.215-223.
6. Reicosky, D.C, W.D. Kemper, G. W. Langdale, C.L. Douglas and P.E. Rasmussen (1995), "Soil Organic Matter Changes Resulting From Tillage and Biomass Production," *Journal of Soil and Water Conservation,* Vol.50, No.3, pp.253-261.
7. Sala, O.E. and J.M. Paruelo (1997), "Ecosystem Services in Grasslands", in G. Daily (Ed) (1997), *Nature's Services: Societal Dependence on Natural Ecosystems,* Island Press, Washington, D.C., U.S.A.
8. Rasmussen, P.E., K.W.T. Goulding, J. R. Brown, P. R. Grace, H.H. Janzen and M. Korschens (1998), "Long Term Agro-ecosystem Experiments: Assessing Agricultural Sustainability and Global Change", *Science,* Vol.282, pp.893-896.
9. Tilman, D. (1998), "The Greening of the Green Revolution", *Nature,* Vol.396, pp.211-212.

10. Smith, K.A. (1999), "After Kyoto Protocol: Can Scientists Make a Useful Contribution?" *Soil Biol. Biochemistry,* Vol.15,pp.71-75.
11. RobertM., J. Antoine and F. Nachtergaele (2001), *Carbon Sequestration in soils, Proposal for Land Management in Arid Areas of the Tropics,* AGLL, Food and Agriculture Organization of the United Nations, Rome, Italy.
12. Kramer, S.B.; J.P. Reganold; J.D. Glover; B.J.M. Bohannan H. A. mooney (2006), " Reduced Nitrate Leaching and Enhanced Denitrifier Activity and Efficiency in Organically Fertilised Soils" *Proceedings of the National Academy of Sciences of the USA.,* Vol. 103, pp. 4522-4527
13. Nemecek, T; O. Hugnenin. Elie, D. Dubois and G. Gailord (2005) "Okobilanzierung von anbausystemen im schweizericschen Acker – und futterbau", *Schriftenreihe der* FAL, 58 FAL Reckenholz, Zurich
14. Regonald, j.P,; L.F. Elliot and Y.L. Unger (1987), Long-Term Effects of Organic and Conventional Farming on Soil Erosion", *Nature,* Vl.330, pp.370-372
15. Siegrist, S., D. Staub, L. Pfiffner and P. Mader (1998) "Does Organic Agriculture Reduce Soil Erodibility? The Results of a Long-Term Field Study on Losses in Switzerland," *Agriculture, Ecosystems and Environment,* Vol.69, pp. 253-264.
16. Niggli, U., A. Fliebach, P. Hepperly, J. hanson, D. Douds and R. Seidel (2009), "Low Greenhouse Gas Agriculture: Mitigation and Adoption Potential of Sustainable Farming System", *Food and Agriculture Organization, Review – 2,* pp.1-22.
17. Mader, P., A. Fliebach, D. Dubois, L. Gunst, P. Fried and U. Niggli (2002), "Soil Fertility and Biodiversity in Organic Farming", *Science,* Vol.296,pp.1694-1697.
18. Pimentel, D., P. Hepperly, J. Hanson, D. Douds and R. Seidel (2005), "Environmental, Energetic and Economic Comparisons of Organic and Conventional Farming Systems", *Bioscience,* Vol.55 pp.573-582.
19. Op. cit
20. Fliessbach, A. and P. Mader (2000), "Microbial Biomass and Size-Density Fractions Differ Between Soils or Organic and Conventional Agriculture Systems", *Soil Biol. Biochemistry,* Vol.32,pp. 757-768.
21. *Op. cit.*
22. *Op. cit.*
23. No single definition of synthetic exists, although the various material lists of allowed and prohibited inputs for organic production, developed in different countries and by different certification programmes, are fairly consistent, reflecting an implicit agreement on a definition. The few legal definitions of 'synthetic' reflect the common understanding of the term in organic trade.
24. Crop rotation is the practice of alternating crops grown on a specific field in a planned pattern or sequence in successive crop years. Organic certification programmes require 'soil building' crop rotations, meaning that they must be specifically designed to steadily improve soil filth and fertility while reducing nitrate leaching, weed, pest and disease problems. IFOAM, for example,

recommends specific rotations that include legumes and requires the rotation of non-perennial crops "in a manner that minimises pressure from insects, weeds, diseases and other pests, while maintaining or increasing soil, organic matter, fertility, microbial activity and general soil health." Under limited cropping conditions (e.g., mushrooms, perennials) crop rotations may not be applicable; in such cases other methods that contribute to soil fertility may be required by certification programmes.

25. 'Natural' is commonly understood as anything that is non-synthetic.
26. UNDP (1992), Benefits of Diversity: An Incentive towards Sustainable Agriculture, *United Nations Development Programme*, New York
27. Lampkin N H (1994) "Economics of organic farming in Britain" in The economics of organic farming – An international perspective (ed) by Lampkin N.H and Padel S., *CAB International Publishers*
28. Statistics and Emerging Trends, 2010, *The World of Organic Agriculture* – IFOAM and FiBL, Frick.
29. The term 'organically managed land' etc. refers to certified organic agriculture and includes both the certified in conversion areas and the certified fully converted areas.
30. Wyss E.,H. Luka,L. Pfiffner,C. Schlatter,G. Uehlinger,C. Daniel "Approaches to Pest Management in Arganic Agriculture: a case study in European apple orchards" Paper presented at a symposium entitled *"IPM in Organic Systems"*, XXII International Congress of Entomology, Brisbane, Australia, 16 August 2004, available on the world wide web: http://www.organic-research.com/
31. Yadav C.P.S., Harimohan Gupta, Dr. R. S. Sharma, Organic Farming and Food Security: A Model for India, Organic Farming Association of India, 2010.
32. Kurma Charyulu D and Subho Biswas (2010), "Economics and Efficiency of Organic Farming vis-à-vis Conventional Farming in India" *Working Paper No. 2010-04-03*, CMA, IIM Ahmadabad, April 2010
33. Prasad, R. (1999), Organic farming *vis-à-vis* modern agriculture *Curr. Sci.*, 1999, 77, 38–43.

References

Arrouays, D. and P.Pelissier (1994), "Changes in Carbon Storage in Temperate Humic Soils After Forest Clearing and Continuous Corn Cropping in France", *Plant Soil*, Vol.160, pp.215-223.

Cole, C.V.; J. Duxbury, J. Freney, O. Heinemeyer, K. Minami, A. Mosier, K. Paustin, N. Rosenberg; N. Sampson, D. Sauerbeck and Q. Zaho (1997), "Global Estimates of Potential Mitigation of Greenhouse Gas Emissions by Griculture," *Nut Cycl Agroecosyst*, Vol. 49, pp. 221-228.

Economic Survey 2011, Planning Commission, Government of India

Fliessbach, A. and P. Mader (2000), "Microbial Biomass and Size-Density Fractions Differ Between Soils or Organic and Conventional Agriculture Systems", *Soil Biol. Biochemistry*, Vol.32,pp. 757-768.

Joshi. P.K., (2010) "Conservation Agriculture: An Overview", *Indian Journal of Agricultural Economics,* Vol.66, No.1 pp.53-63.

Kramer, S.B.; J.P. Reganold; J.D. Glover; B.J.M. Bohannan H. A. mooney (2006), " Reduced Nitrate Leaching and Enhanced Denitrifier Activity and Efficiency in Organically Fertilised Soils" *Proceedings of the National Academy of Sciences of the USA.,* Vol. 103, pp. 4522-4527

Kurma Charyulu D and Subho Biswas (2010), "Economics and Efficiency of Organic Farming vis-à-vis Conventional Farming in India" *Working Paper No. 2010-04-03,* CMA, IIM Ahmadabad, April 2010

Lampkin N H (1994) "Economics of organic farming in Britain" in The economics of organic farming - An international perspective (ed) by Lampkin N.H and Padel S., *CAB International Publishers*

Mader, P., A. Fliebach, D. Dubois, L. Gunst, P. Fried and U. Niggli (2002), "Soil Fertility and Biodiversity in Organic Farming", *Science,* Vol.296,pp.1694-1697.

Mohana Rao. L.K (2011) Detailed discussion on the general economic development of India in the recent past, budget meet 2011 held at Dept. of Economics, Andhra University, Visakhapatnam

Nemecek, T; O. Hugnenin. Elie, D. Dubois and G. Gailord (2005) "Okobilanzierung von anbausystemen im schweizericschen Acker – und futterbau", *Schriftenreihe der* FAL, 58 FAL Reckenholz, Zurich

Niggli, U., A. Fliebach, P. Hepperly, J. hanson, D. Douds and R. Seidel (2009), "Low Greenhouse Gas Agriculture: Mitigation and Adoption Potential of Sustainable Farming System", *Food and Agriculture Organization, Review – 2,* pp.1-22.

Pimentel,D., P. Hepperly, J. Hanson, D. Douds and R. Seidel (2005), "Environmental, Energetic and Economic Comparisons of Organic and Conventional Farming Systems", *Bioscience,* Vol.55 pp.573-582.

Prasad, R. (1999), Organic farming *vis-à-vis* modern agriculture *Curr. Sci.,* 1999, 77, 38–43.

Pretty, Jules and Ball Andrew (2001), Agricultural Influences on Carbon Emissions and Sequestration: A Review of Evidence and the emerging Trading Options, *Occasional Paper, Centre for Environment and Society and Department of Biological Sciences,* University of Essex, U.K.

Rasmussen, P.E., K.W.T. Goulding, J. R. Brown, P. R. Grace, H.H. Janzen and M. Korschens (1998), "Long Term Agro-ecosystem Experiments: Assessing Agricultural Sustainability and Global Change", *Science,* Vol.282, pp.893-896.

Regonald, J.P, L.F. Elliot and Y.L. Unger (1987), Long-Term Effects of Organic and Conventional Farming on Soil Erosion", *Nature,* Vl.330, pp.370-372

Reicosky, D.C, W.D. Kemper, G. W. Langdale, C.L. Douglas and P.E. Rasmussen (1995), "Soil Organic Matter Changes Resulting From Tillage and Biomass Production," *Journal of Soil and Water Conservation,* Vol.50, No.3, pp.253-261.

RobertM., J. Antoine and F. Nachtergaele (2001), *Carbon Sequestration in soils, Proposal for Land Management in Arid Areas of the Tropics,* AGLL, Food and Agriculture Organization of the United Nations, Rome, Italy.

Sala, O.E. and J.M. Paruelo (1997), "Ecosystem Services in Grasslands", in G. Daily (Ed) (1997), *Nature's Services: Societal Dependence on Natural Ecosystems*, Island Press, Washington, D.C., U.S.A.

Siegrist, S., D. Staub, L. Pfiffner and P. Mader (1998) "Does Organic Agriculture Reduce Soil Erodibility? The Results of a Long-Term Field Study on Losses in Switzerland," *Agriculture, Ecosystems and Environment,* Vol.69, pp. 253-264.

Smith, K.A. (1999), "After Kyoto Protocol: Can Scientists Make a Useful Contribution?" *Soil Biol. Biochemistry,* Vol.15,pp.71-75.

Statistics and Emerging Trends, 2010, *The World of Organic Agriculture* - IFOAM and FiBL, Frick.

The Organic Standard and The Agricultural and Processed Food Products Export Development Authority (APEDA) 2010.

Tilman, D. (1998), "The Greening of the Green Revolution", *Nature,* Vol.396, pp.211-212.

UNDP (1992), Benefits of Diversity: An Incentive towards Sustainable Agriculture, *United Nations Development Programme,* New York

Wyss E.,H. Luka,L. Pfiffner,C. Schlatter,G. Uehlinger,C. Daniel "Approaches to Pest Management in Arganic Agriculture: a case study in European apple orchards" Paper presented at a symposium entitled *"IPM in Organic Systems",* XXII International Congress of Entomology, Brisbane, Australia, 16 August 2004,

Yadav C.P.S., Harimohan Gupta, Dr. R. S. Sharma, *Organic Farming and Food Security: A Model for India,* Organic Farming Association of India, 2010.

Chapter 2
Review of Literature

Reviewing the existing literature on any proposed research is very important for any researcher to have a clear-cut idea on the Problem and it is very useful in analyzing and interpreting the data for drawing some meaningful conclusions. With this view, in this Chapter, an attempt has been made to present the studies conducted by various researchers, both at national and international levels on various issues relating to organic farming.

Wyss *et. al.*,[1] (2004) traced the history of organic farming in Europe and pointed out to different strategies to be adopted.

A Study conducted by Stolze Matthias and Nicolas Lampkin (2009)[2] concluded that since the mid 1980s, organic farming has become the focus of significant attention from policy-makers, consumers, environmentalists and farmers in Europe and state institutions have become increasingly involved in regulating and supporting the organic sector. Reflecting on the multiple goals for organic farming and for agricultural policy, the Study pointed out a varied and complex range of policy measures that have been developed and implemented to support the organic sector. However, the study contained that balancing societal and consumer/market goals and balancing institutional and private stakeholder interests in the organic sector pose challenges for policy-making both in the dimension of policies and the dimension of politics.

Anderson, J.C., *et al.*, (2006)[3] conducted a study and concluded that organic food was perceived by respondents to be, in general, a healthier alternative to "regular food," including its effect on appearance and resulting from higher nutrient levels.

The FAO Committee on Agriculture (COAG) (1999)[4], which met in Rome in January 1999 reported that although still only a small industry, organic agriculture is gaining of growing importance in the agricultural sector of many countries, and reported higher opportunities for export markets, irrespective of their stage of development. The Report has exhaustively discussed the details and they are summarized below:

Since the demand for a variety of organic products is high, many developing countries have started to tap those lucrative export markets for organically grown products - for example, tropical fruit to the European baby food industry, Zimbabwe herbs to South Africa, African cotton to the E U, and Chinese tea to the Netherlands and Soybeans to Japan.

Typically, organic exports are sold at impressive premia, often at prices 20 per cent higher than identical products produced on non-organic farms. The ultimate profitability of organic farming varies. Only a limited number of studies have assessed its long-term prospects. Farmers and agribusinesses seek to sell their products in developed countries usually hiring an organic certification agency to annually inspect and confirm that they adhere to the standards established by various trading partners. The cost for this service can be expensive, especially since few developing countries have certification organizations of their own.

Typically, farmers experience some loss in yields after discarding synthetic inputs and converting their operations to organic production. Sometimes it may take years to restore the ecosystem to the point where organic production is possible. In these cases other sustainable approaches that allow judicious use of synthetic chemicals may be more suitable start-up options. One strategy involves converting farms to organic production "in installments", so that the entire operation is not put at risk.

Most studies have found that organic agriculture requires significantly greater labour input compared to conventional farms. Therefore, the diversification of crops typically found on organic farms, with their different planting and harvesting schedules, may distribute labour demand more evenly, which could help stabilize employment. As in all agricultural systems, diversity in production increases income-generating opportunities and can, as in the case of fruits, supply the essential health-protecting minerals and vitamins for the family diet. It also spreads the risks of failure over a wide range of crops.

Nevertheless, organic farmers face huge uncertainties. Some studies noted that 73 per cent of North American organic farmers reported lack of information on organic conversion, as the extension personnel have inadequate training in organic methods and as they sometimes discourage farmers to adopt organic farming. Furthermore, institutional support in developing countries is found to be scarce.

Land tenure is also critical to the adoption of organic agriculture. It is highly unlikely that tenant farmers would invest the necessary labour, and sustain the difficult conversion period, without some guarantee of access to the land in later years, when the benefits of organic production emerge.

Most organic farmers are motivated by more than economic objectives - their aim is to optimize land, animal, and plant interactions, preserve natural nutrient and energy flows, and enhance biodiversity, all of which contribute to sustainable agriculture. Their use of crop rotations, organic manure and mulches improves soil structure and encourages development of a vigorous population of soil micro-organisms. Mixed and relay cropping provide a more continuous soil cover and thus a shorter period when the soil is fully exposed to the erosive power of the rain, wind and sun.

Organic farmers also employ natural pest controls - e.g. biological control, plants with pest control properties - rather than synthetic pesticides which, when misused, are known to kill beneficial organisms, cause pest resistance and often pollute water and land. Reduction in the use of toxic synthetic pesticides, which poison an estimated three million people each year, should lead to improved health of farm families.

Finally, eliminating the use of synthetic nitrogenous fertilisers greatly lowers the risks of nitrogen contamination of water, while crop rotation is a widely used method of fertility maintenance and pest and disease control. Most certification programmes restrict the use of mineral fertilisers, which may instead be necessary to supplement organic manure produced on the farm. However, natural and organic fertilizers from outside the farm may also be used and crop rotations encourage a diversity of food crops, fodder and under-utilized plants which, in addition to improving overall farm production and fertility, may assist in the on-farm conservation of plant genetic resources.

Delate Kathleen *et al.*, (2003)[5] stated that as transition to organic production and increasing public demand for organic products attracts premium prices for the certified organic farmer, it makes the conventional farmers to consider going to organic way. They assessed the agro-ecosystem performance of farms during the three-year transition it takes to switch from conventional to certified organic grain production. Their Study found that organic grain crops can be successfully produced in the third year of transition and that additional economic benefits can be derived from expanded crop rotation. Their Study tested the hypothesis that organic systems relying on locally derived inputs are capable of providing stable yields, while maintaining soil quality and plant protection compared with conventional systems with less diverse crop rotations and greater levels of external, fossil-fuel based inputs.

After a 21-year study, the Swiss scientists Mader Paul, *et al.*, (2002)[6], have given a ringing endorsement to organic farming methods. They found that organic yields were on average 20 per cent lower than those from conventional agriculture. But the ecological benefits are more and the organic crops proved more efficient users of energy and other resources. Their study concluded that organic farming is a viable alternative to conventional ways of farming.

Miller P.R. *et al.*, (2008)[7] conducted a study to compare several transitional crop productivity and soil nutrient status among diversified NT (Not Tillage) and ORG (Organic Diversified) cropping systems in Montana. Studying simultaneous transition for a four years period to diversified NT and ORG cropping systems was instructive for increased sustainability.

Dimitri Carolyn *et al.*,(2004)[8] summarized growth patterns in the U.S. organic sector in recent years, by market category, and described various research, regulatory, and other ongoing programs on organic agriculture in the U.S. Department of Agriculture.

Gu¨ndog¢mus Erdemir (2006)[9] compared the energy-use in apricot production both on organic and conventional farms in Turkey in terms of energy ratio, benefit/cost ratio and amount of renewable energy use. The total energy requirement under organic apricot farming was 13,779.35 MJ ha^{-1}, whereas 22,811.68 MJ ha^{-1} was consumed under conventional apricot farming, i.e. 38% higher energy input was used on conventional apricot farming than the use on organic farms. The energy ratios of 2.22 and 1.45 were achieved under the organic and conventional farming systems, respectively.

Abouleish Helmy (2007)[10] in his study entitled "Organic agriculture and food Utilisation - an Egyptian case study" concluded that the quality of drinking water will improve further with an expected expansion of organic agriculture and organic agriculture enables ecosystems to better adjust to the effects of climate change and has a major potential for reducing agricultural greenhouse and other gas emissions. Further, he mentioned the results of Shame Heaton, which found that the organic farmer contains fewer pesticides. If at all used as they degrade quickly and rarely leave any residue on organic food. As a result of introducing organic agriculture in Egypt's cotton sector, the annual amount of pesticides-use was reduced from 30,000 tons in the early 1990s to around 3,000 tons by 2007. This is the most remarkable contribution of organic agriculture to food quality and health in Egypt. As far as the food quality is concerned, the Study revealed that organic produce contains more nutrients: all nutrients on average are higher in organic produce, this is particularly significant in the case of vitamin C, magnesium, iron, phosphorus etc., these naturally occur in plants and protect them from disease and pest.

Kassie, Menale *et al.*, (2008)[11] in their study stated that Organic farming practices, in as far as they rely on local or farm renewable resources, present desirable options for enhancing agricultural productivity for resource-constrained farmers in developing countries particularly in Ethiopia. Results of their Study underscored the importance of encouraging resource-constrained farmers in developing countries to adopt organic farming practices, especially, since they enable farmers to reduce production costs, provide environmental benefits, and as the results confirm to enhance crop productivity.

Reganold, JP *et al.*, (2001)[12] concluded that escalating production costs, heavy reliance on non-renewable resources, reduced biodiversity, water contamination, chemical residues in food, soil degradation and health risks to farm workers handling pesticides all bring into question the sustainability of conventional farming systems of apple production for 1994-99. It has been claimed, however, that organic farming systems are less efficient, and produce half the yields of conventional farming systems. Nevertheless, organic farming became one of the fastest growing segments of US and European agriculture during the 1990s. Integrated farming, using a combination of organic and conventional techniques, has been successfully adopted on a wide scale in Europe. The organic and integrated systems had higher soil quality and potentially lower negative environmental impact than the conventional system. When compared with the conventional and integrated systems, the organic system produced sweeter apples, higher profitability and greater energy efficiency. The results, further indicated that the organic system ranked first in environmental and economic sustainability, the integrated system second and the conventional system last.

Pimentel David (2005)[13] concluded that Organic Farming offers real advantages for such crops as corn and soybean and analyzed the environmental, energy and economic costs and benefits of growing soybeans and corn organically versus conventionally. Their Study is a review of the Rodale Institute Farming Systems Trial, the longest running comparison of organic vs. conventional farming in the United States.

Anand Raj Daniel *et al.*, (2005)[14] in their study concluded that in 2004 organic cotton yielded generally on par with conventional cotton. In the case of organic cotton grown on fields that came out of a short term fallow, yields were higher than yields of conventional cotton. Profitability of organic cotton was significantly higher than that of conventional cotton, the contributing factor being reduced expenditure on pest control management (PCM).

Swezey S L *et al.*, (2004)[15] in their study compared three different cotton production strategies in field-sized replicates in the Northern San Joaquin Valley (NSJV), California in the USA, for 1996 - 2001. Cotton production treatments included certified organic, conventionally grown and supervised integrated pest management (IPM) strategies. Lower quantities of insecticide were used in organic and IPM fields than in conventional fields. This cost differential between organic and conventional cotton was primarily due to greater hand-weeding costs and lower yields in organic cotton. Yields were 2.1, 2.7, and 2.8 bales/acre, for organic, IPM and conventional treatments, respectively. Low world cotton prices and the lack of premium prices for organic cotton were found to be the primary obstacles for its continued production in the NSJV in USA.

Vangelis Tzouvelekas *et al.*, (2001)[16] using the recent advances in the stochastic production frontier framework, presented an empirical analysis of technical, allocative and economic efficiency of a sample of organic and conventional cotton farms located in Greece, and suggested that both farm types in the sample examined are technically, allocatively and economically inefficient. Farmers' age and education and farm size were found to be important factors in explaining differentials in efficiency estimates. In comparative terms, organic farms exhibited lower efficiency scores *vis-à-vis* their conventional counterparts in terms of technical and economic efficiency; regarding allocative efficiency both farm types are almost equally inefficient. Low efficiency scores in both types of farming may be attributed to the respective intervention policies of the last 20 years.

Lesjak Heli Annika (2008)[17] argued, based on 16 distinct assessment criteria, during 1960 and 1994 that the growth of organic farming correlates with the past support policy decisions. The recent direct organic farming payments are of no importance, but on the extent to which the past policies focused on rural development. Building on the OECD Positive Policy Principles, the Study assessed the support policies of Austria, Finland and the EU.

Posner Joshua L *et al.*, (2008)[18] observed that during the last half-century, agriculture in the upper U.S. Midwest has changed from limited-input, integrated grain–livestock systems to primarily high-input specialized livestock or grain systems. This trend has spawned a debate regarding which of cropping systems is more sustainable and led to the question: "can the diverse, low-input cropping systems like organic systems be as productive as the conventional systems?" To answer this question, they compared six cropping systems ranging from diverse, organic systems to less diverse conventional systems at two sites in southern Wisconsin. The results of their 13 years Study at one location and eight years Study in another showed that: (i) organic forage crops can yield both as much dry matter as their conventional counterparts and with quality, sufficient to produce as much milk; and (ii) Several crops can produce 90 per cent as well as their conventionally managed counterparts. Combining with other controlled data, they found that weed control was a problem, resulting in lower yields, Finally, their findings indicate that diverse, low-input cropping systems can be as productive per unit of land as conventional systems.

Pluke Richard and Amy Guptill (2004)[19] studied social, ecological and farming system constraints to organic crop protection in Puerto Rico using a linear programming model, for systems analysing reasons behind this anomaly. Many of the reasons lied in the historical marginalization of agriculture. Without a strong agricultural sector, Puerto Rico's mixed economic developments of the 20th century and the U.S.'s response to the rising poverty levels only exacerbated dependency. Cheap imports, food stamps and a comprehensive agricultural incentives program virtually ensured that farmers are not in a position to develop a significant organic farming sector. This is particularly true of the central mountainous region where most of the island's smallest farms are found. The linear programming model indicated that labor and poor markets are the biggest constraints to the producers of the central region. Organic crop protection strategies can often more be labor-intensive and, without a strong, dependable market, most of the farmers would not invest in the additional labor needed to develop organic production. On a more positive note, many of the crops grown in the central region of Puerto Rico are managed without pesticides. This is in part to do with producers choosing crops that have low labor requirements.

Wood Richard *et al.,* (2005)[20] examined the causes for environmental impacts in Australia that range from local through global in scale. They assessed on farm and indirect energy consumption, land disturbance, water use, employment, and emissions of greenhouse gases, of organic and conventional farming in Australia. While organic farming may be argued to be superior to conventional farming on the basis of local impacts, it is not often clear how organic farming performs relative to conventional farming in terms of wider, global impacts. However, they found that the indirect contributions for all factors are much higher for the conventional farms. Showing that indirect effects must be taken into account in the consideration of the environmental consequences of farming, in particular for energy use and greenhouse gas emissions, where the majority of impacts usually occur off-farm. Finally, subject to yield uncertainties for organic versus conventional farming, from the sample study, they concluded that in addition to their local benefits, organic farming approaches can reduce the total water, energy and greenhouse gases involved in food production.

Acs S *et al.*, (2006)[21] opined that organic farming is more profitable than conventional farming. However, in reality not many farmers convert to organic farming. Policy makers and farmers do not have a clear insight into factors which hamper or stimulate the conversion to organic farming. They as such developed a dynamic linear programming model to analyse the effects of different limiting factors on the conversion process of farms over time. The Model developed for a typical arable farm in Netherlands central clay region, is based on two static liner programming models (conventional and organic), with an objective to maximise the net present value over a 10-year planning horizon. The results of the analysis of a basic scenario showed that conversion to organic farming is more profitable than staying conventional.

Kirchmann Holger *et al.*, (2007)[22] conducted an 18-yr field study to compare organic and conventional cropping on a highly P and K depleted soil in southern Sweden that had not received any inorganic fertilizers (or pesticides) since the mid-1940s. The major agronomic management differences between five systems viz. (i) growth of legumes every second year and use of legumes as cover crops in the organic rotation; (ii) application of P in the organic system at higher rates than for the conventional system; (iii) exclusion of oilseed rape (*Brassica napus* L.) from the organic system but inclusion of potato (*Solanum tuberosum* L.); (iv) frequent mechanical weeding in the organic system; and (v) use of solid manure in the organic and liquid manure in the conventional system were found to be that long-term use efficiency of P was lower in the organic system (seven per cent) than in the conventional system (36 per cent). These results showed that yield and soil fertility are superior in conventional cropping systems under cold-temperate conditions.

Prasad, R.,(1999)[23] in his study on organic farming *vis-à-vis* modern agriculture concluded that organic farming, as in the modern context, was practised in India only on Ha. 4800 in 2003 and the produce exported was valued at about ₹ 89 crores, which is only 0.80 per cent of the current global market. Among the field crops, only Basmati rice, cotton and sesame were exported. Cotton and sesame are mostly grown under rain-fed/dry-land agricultural conditions, and it should not be difficult to grow these crops using organic manure. Cotton is the largest consumer of insecticides and real serious efforts will prevent their use to guarantee organically produced cotton by demarcating areas and restricting pest control to neem and other botanical insecticides and bio-pesticides. Basmati rice is grown in the north in the 'rice–wheat cropping system' belt, where large amounts of fertilizers are used. Here again, areas need to be demarcated. Reasonable price guarantee can do the trick as yield levels in organic manure fields are likely to be lower. Similar is the case with fruits and vegetables.

He suggested that use of organic matter improves soil structure and increases water-holding capacity, which is important under dry farming conditions and assures a regular supply of micronutrients. Nevertheless, availability of macronutrients from organic manure is not as fast as from chemical fertilizers, because it depends upon the rate of their decomposition. However, he contained that myths such as better taste, improved quality and higher nutritive value generally attached with organically produced foods have been argued and found to lack a scientific basis. Nevertheless, market for organically produced foods is on the increase. India can greatly benefit from the export of organic foods, but needs to seriously devote attention to market intelligence regarding which products to grow, where to sell, distribution channels, competition, market access, etc. He suggested that pre-harvest prices should be announced, so that farmers do not suffer when the produce is ready as organic farming is a market demand oriented, highly specialized small sector of Indian agriculture, which if well planned and executed can become an important foreign-exchange earner for the country and money-spinner for the farmers.

Singh Y. V *et al.*, (2007)[24] in one of their studies which is mostly on agronomic practices of organic farming in India observed that management of soil organic matter is critical to maintain a productive organic farming system. No one source of nutrient usually suffices to maintain productivity and quality control in organic system. In addition, the inputs to supplement nutrient availability are often not uniform presenting additional challenges in meeting the nutrient requirement of crops in organic systems. With this concept, a field experiment was conducted at the research farm of Indian Agricultural Research Institute (IARI), New Delhi, during 2003-06 in rice-wheat-green gram cropping system. An interesting observation recorded was that there was no serious attack of any insect pest or disease in organically grown crop. Soil microbial population enhanced due to the application of organic amendments in comparison to absolute control as well as recommended fertilizer application that in turn resulted in a notable enhancement in soil dehydrogenase and phosphatase enzyme activity. However, to meet the ever-growing food-grain demands of the country, which is estimated at 294 million tonnes per annum by 2020, the mainstream of Indian agriculture has to depend on modern agricultural inputs, such as chemical fertilizer and pesticides. Nevertheless, their restrained and efficient use is important. As regards plant nutrient needs in modern agriculture, integrated nutrient supply is the key for sustainable Indian agriculture.

Kurma Charyulu, D and Subho Biswas(2010)[25] focused mainly on the issues like economics and efficiency of organic farming *vis- à- vis* conventional farming in India. Four states of Gujarat, Maharashtra, Punjab and U.P were purposively selected for the Study. Similarly, four major crops i.e., cotton, sugarcane, paddy and wheat were chosen for the comparison. A Model based nonparametric Data Envelopment Analysis (DEA) was used for analyzing the efficiency of the farming systems. The results showed a mixed response. Overall, it is concluded that the unit costs of production is lower in organic farming in case of cotton and sugarcane, whereas the same is lower in conventional farming for paddy and wheat. The DEA efficiency analysis conducted on four different crops indicated that the efficiency levels are lower in organic farming compared to conventional farming, relative to their production frontiers. The results concluded that there is ample scope for increasing the efficiency under organic farms.

In another study Kurma Charyulu, D and Subho Biswas (2010)[26] observed that the entire agricultural community is trying to find out an alternative sustainable farming system, which is ecologically sound and economically and socially acceptable. Traditional agricultural practices, which are, based on natural and organic methods of farming offer several effective, feasible and cost effective solutions to most of the basic problems being faced in conventional farming system. National Project on Organic Farming (NPOF) and National Programme for Organic Production (NPOP) same in this direction.

The preceding review of literature clearly has brought to the fore that the spread of Organic Farming is relatively higher in advanced countries like USA, Switzerland and Western Europe and it is gaining momentum in developing countries like India. It also revealed that Organic Farming is beneficial to the human and other living beings by way of providing qualitative food products, protecting environment, etc. However, there is inconclusive evidence on the economic gaining/ profitability of Organic farming and there exists a dearth of studies on this aspect in the Indian context. Further, except the pioneering works at the CMA, IIM, Ahmadabad, which focused their attention on the Northern and Western parts of India, on paddy, wheat, sugarcane and cotton and on the efficiency of inputs used in organic farming and conventional farming and another peripheral study by Prasad which studied several comparative aspects of organic farming and conventional farming, while the Study of Singh et al. has been mostly on agronomic one, no researcher in India has so far examined location-specific and crop-specific aspects relating to commodities of organic farming and conventional farming covering in a State. Hence, a comprehensive study dealing with economics of organic farming and conventional farming covering different agro-climatic conditions is felt necessary. As such, the present Study addressed itself to fill in this gap by examining the Economics of Organic Farming vis-à-vis Conventional Farming in A.P. covering cereals, pulses and oil-seeds in East Godavari, Mahabubnagar and Anantapur.

Notes

1. Wyss E.,H. Luka,L. Pfiffner,C. Schlatter,G. Uehlinger,C. Daniel "Approaches to Pest Management in Organic Agriculture: A Case Study in European Apple Orchards" Paper presented at a symposium entitled *"IPM in Organic Systems"*, XXII International Congress of Entomology, Brisbane, Australia, 16 August 2004, available on the world wide web: http://www.organic-research.com/
2. 2 Stolze Matthias and Nicolas Lampkin (2009) "Policy for organic farming: Rationale and concepts " Published by "Elsevier ". 0306-9192/$ - see front matter 2009 doi:10.1016/j.foodpol.2009.03.005 and also available on the world wide web: http://www.sciencedirect.com/science/article/B6VCB-4W2M6V2-2/2/d9fe87e79f7605d0f2d36223ff57298e
3. 3 Anderson, J.C., Wachenheim, C.J., & Lesch, W.C. (2006). "Perceptions of Genetically Modified and Organic Foods and Processes". *AgBioForum,* 9(3), 180-194. available on the world wide web: http://www.agbioforum.org.
4. 4 Demand for organic products has created new export opportunities for the developing world, Spotlight / 1999, a Report to the *FAO Committee on Agriculture (COAG),* which met in Rome on 25-26 January 1999. available on the world wide web : http://www.fao.org/ag/magazine/9901sp3.htm.
5. 5 \Delate Kathleen and Cynthia A., "Organic Production Works", Online, *Institute of Science in Society,* available on the world wide web: http://lists.ifas.ufl.edu/cgi-bin/wa.exe
6. 6 **Mader Paul, et. al. 2002,** "Soil Fertility and Biodiversity in Organic Farming", *Science* 31 May 2002: Vol. 296. no. 5573, pp. 1694 – 1697, DOI: 10.1126/science.1071148, available on the world wide web: http://www.sciencemag.org/cgi/content/abstract/296/5573/1694
7. 7 Miller Perry R., David E. Buschena, Clain A. Jones and Jeffrey A. Holmes (2008), "Transition from Intensive Tillage to No-Tillage and Organic Diversified Annual Cropping Systems", Published n *"Agron J"* 100:591-599 (2008) DOI: 10.2134/agronj2007.0190, available on the world wide web: http://agron.scijournals. org/cgi /content/abstract/100/3/591
8. Dimitri Carolyn and Catherine Greene, Recent Growth Patterns in the U.S. Organic Foods Market U.S. Department of Agriculture, Economic Research Service, Market and Trade Economics Division and Resource Economics Division. Agriculture Information Bulletin Number 777. available on world wide web: http://www.ers.usda.gov/publications/aib777/aib777a.pdf
9. Gu¨ndog¢mus Erdemir (2006) "Energy use on organic farming: A comparative analysis on organic versus conventional apricot production on small holdings

in Turkey" Published in *"ELSEVIER"*. and also available on the world wide web http:// doi:10.1016/j.enconman.2006.01.001

10. Abouleish Helmy (2007) "Organic agriculture and food Utilisation - an Egyptian case study" Managing Director, SEKEM Group, Egypt, available on the world wide web :ftp://ftp.fao. org/paia /organicag/ofs/07 -Abouleish.ppt
11. Kassie, Menale, Zikhali, Precious, Pender, John and Köhlin, Gunnar(2008) "Organic Farming Technologies and Agricultural Productivity: The case of Semi-Arid Ethiopia" Paper provided by Göteborg University, Department of Economics in its series Working Papers in Economics with number 334, 2008, available on the world wide web:http://ideas.repec.org/p/hhs/gunwpe/0334.html.
12. Reganold, JP., Glover, JD., Andrews, PK ., and Hinman, HR, (2001), "Sustainability of Three Apple Production Systems" *Nature* [Nature]. Vol. 410, no. 6831, pp. 926-930. 19 Apr 2001.
13. Pimentel David "Organic Farming Produces Same Corn and Soybean Yields as Conventional Farms, but Consumes Less Energy and No Pesticides", *Bioscience,* Vol. 55:7, available on the world wide web: http://www.news.cornell.edu/stories/July05/organic.farm.vs.other.ssl.html
14. Anand Raj Daniel, K. Sridhar, Arun Ambatipudi, H. Lanting, & S. Brenchandran, Second "International Symposium on Biological Control of Arthropods", available on the world wide web: http://www.bugwood.org / arthropod2005/ vol1/6c.pdf
15. Swezey Sean L.,Polly Goldman,Janet Bryer,Diego Nieto "Comparison Between Organic, Conventional, and IPM Cotton in the Northern San Joaquin Valley, California", Paper presented at a symposium entitled *"IPM in Organic Systems",* XXII International Congress of Entomology, Brisbane, Australia, 16 August 2004 available on the world wide web: http://www.organic-research.com/
16. Tzouvelekas Vangelis, Christos J. Pantzios, and Christos Fotopoulos, "Economic Efficiency in Organic Farming: Evidence From Cotton Farms in Viotia, Greece" *Journal of Agricultural &Applied Economics.,* Volume 33, April, 2001, Issue: 1, Pp: 35-48, available on the world wide web: http://ideas.repec.org/a/jaa/jagape/v33y2001i1p35-48.html#abstract
17. Lesjak Heli Annika (2008) "Explaining organic farming through past policies: comparing support policies of the EU, Austria and Finland" *Journal of Cleaner Production 16 (2008) 1- 11* and also available on world wide web: http://doi:10.1016/j.jclepro.2006.06.005
18. Posnera Joshua L.,, Jon O. Baldock and Janet L. Hedtcke (2008), "Organic and Conventional Production Systems in the Wisconsin Integrated Cropping Systems Trials: I". Productivity 1990–2002, Published in *"Agron J"* 100:253-260, 2008 American Society of Agronomy, 677 S. , Segoe Rd., Madison, WI 53711 USA, available on the world wide web: http://agron.scijournals.org/cgi /content/abstract/100/2/253
19. Pluke Richard,Amy Guptill "The Social, Ecological and Farming System

Constraints on Organic Crop Protection in Puerto Rico" Paper presented at a symposium entitled *"IPM in Organic Systems"*, XXII International Congress of Entomology, Brisbane, Australia, 16 August 2004., available on the world wide web : http://www.organic-research.com/

20. Wood Richard, Manfred Lenzen, Christopher Dey, Sven Lundie (2005) "A Comparative Study of Some Environmental Impacts of Conventional And organic Farming in Australia" Published in *"ELSEVIER"*. and also available on the world wide web: http:// doi:10.1016/j.agsy.2005.09.007

21 Acs S., P.B.M. Berentsen, R.B.M. Huirne (2006) "Conversion to organic arable farming in The Netherlands: A dynamic linear programming analysis" Published in *"ELSEVIER "* and also available on the world wide web: http://doi:10.1016/j.agsy.2006.11.002.

22. Kirchmann Holger, Lars Bergström, Thomas Kätterer, Lennart Mattsson and Sven Gesslein (2007), "Comparison of Long-Term Organic and Conventional Crop–Livestock Systems on a Previously Nutrient-Depleted Soil in Sweden", Published in *Agron* J 99:960-972 (2007), DOI: 10.2134/agronj2006.0061© 2007 American Society of Agronomy, 677 S. Segoe Rd., Madison, WI 53711 USA, available on the world wide web: http://agron.scijournals. org/cgi/content/abstract/ 99/4/960

23. Prasad, R. (1999), Organic farming *vis-à-vis* modern agriculture *Curr. Sci.*, 1999, 77, 38–43.

24. Singh Y. V., B. V. Singh, S. Pabbi and P. K. Singh(2007) "Impact of Organic Farming on Yield and Quality of BASMATI Rice and Soil Properties" available on world wide web http://orgprints.org/9783

25. Charyulu D.K. and Subho Biswas "Economics and Efficiency of Organic Farming vis-à-vis Conventional Farming in India" *Working Paper No. 2010-04-03,* CMA, IIM Ahmadabad, April 2010

26. Charyulu D.K. and Subho Biswas "Efficiency of Organic Input Units under NPOF Scheme in India" *Working Paper No. 2010-04-01,* CMA, IIM Ahmadabad, April 2010.

References

Abouleish Helmy (2007) "Organic agriculture and food Utilisation - an Egyptian case study" Managing Director, SEKEM Group, Egypt, available on the world wide web :ftp://ftp.fao. org/paia /organicag/ofs/07 -Abouleish.ppt

Acs S., P.B.M. Berentsen, R.B.M. Huirne (2006) "Conversion to organic arable farming in The Netherlands: A dynamic linear programming analysis" Published in *"ELSEVIER"* and also available on the world wide web: http://doi:10.1016/j.agsy.2006.11.002

Anand Raj Daniel, K. Sridhar, Arun Ambatipudi, H. Lanting, & S. Brenchandran,(2005)Second "International Symposium on Biological Control of Arthropods", available on the world wide web: http://www.bugwood.org/arthropod2005/ vol1/6c.pdf

Anderson, J.C., Wachenheim, C.J., & Lesch, W.C. (2006). "Perceptions of genetically modified and organic foods and processes". *AgBioForum,* 9(3), 180-194. available on the world wide web: http://www.agbioforum.org

Delate Kathleen and Cynthia A. (2003), "Organic Production Works", Online, *Institute of Science in Society,* available on the world wide web: http://lists.ifas.ufl.edu/cgi-bin/wa.exe?A2=ind0412&L=sanet13

Demand for organic products has created new export opportunities for the developing world, Spotlight / 1999, This article is based on Organic agriculture, a report to the *FAO Committee on Agriculture (COAG),* which met in Rome on 25-26 January 1999.

Dimitri Carolyn and Catherine Greene(, Recent Growth Patterns in the U.S. Organic Foods Market U.S. Department of Agriculture, Economic Research Service, Market and Trade Economics Division and Resource Economics Division. Agriculture Information Bulletin Number 777. available on world wide web: http://www.ers.usda.gov/publications/aib777/aib777a.pdf

Gu¨ndog¢mus Erdemir (2006) "Energy use on organic farming: A comparative analysis on organic versus conventional apricot production on small holdings in Turkey" Published in *"ELSEVIER".* and also available on the world wide web http:// doi:10.1016/j.enconman.2006.01.001

Kassie, Menale, Zikhali, Precious, Pender, John and Köhlin, Gunnar(2008) "Organic Farming Technologies and Agricultural Productivity: The case of Semi-Arid Ethiopia" Paper provided by Göteobrg University, Dept. of Economics in its series Working Papers in Economics with number 334, 2008, available on the world wide web:http://ideas.repec.org/p/hhs/gunwpe/0334.html.

Kirchmann Holger, Lars Bergström, Thomas Kätterer, Lennart Mattsson and Sven Gesslein (2007), "Comparison of Long-Term Organic and Conventional Crop–Livestock Systems on a Previously Nutrient-Depleted Soil in Sweden", Published in *Agron* J 99:960-972 (2007), American Society of Agronomy, 677 S. Segoe Rd., Madison, WI 53711 USA, available on the world wide web: http://agron.scijournals. org/cgi/content/ abstract/ 99/4/960

Kurma Charyulu D and Subho Biswas (2010), "Economics and Efficiency of Organic Farming vis-à-vis Conventional Farming in India" *Working Paper No. 2010-04-03,* CMA, IIM Ahmadabad, April 2010

Kurma Charyulu, D and Subho Biswas (2010), "Efficiency of Organic Input Units under NPOF Scheme in India" *Working Paper No. 2010-04-01,* CMA, IIM Ahmadabad, April 2010.

Lesjak Heli Annika (2008) "Explaining organic farming through past policies: comparing support policies of the EU, Austria and Finland" *Journal of Cleaner Production 16 (2008) 1- 11* and also available on world wide web: http://doi:10.1016/j.jclepro.2006.06.005

Mader Paul, *et. al.* (2002), "Soil Fertility and Biodiversity in Organic Farming", *Science* 31 May 2002: Vol. 296. no. 5573, pp. 1694 – 1697, DOI: 10.1126/science.1071148, available on the world wide web: http://www.sciencemag.org/cgi/content/abstract/296/5573/1694.

Miller Perry R., David E. Buschena, Clain A. Jones and Jeffrey A. Holmes (2008), "Transition from Intensive Tillage to No-Tillage and Organic Diversified Annual Cropping Systems", Published n *"Agron J"* 100:591-599 (2008) DOI: 10.2134/ agronj2007.0190, available on the world wide web: http://agron.scijournals.org/cgi /content/abstract/100/3/591

Pimentel David (2005) "Organic farming produces same corn and soybean yields as conventional farms, but consumes less energy and no pesticides", *Bioscience*, Vol 55:7, available on the world wide web: http://www.news. cornell.edu/stories/July05/organic.farm.vs.other.ssl.html

Pluke Richard,Amy Guptill (2004) "The Social, Ecological and Farming System Constraints on Organic Crop Protection in Puerto Rico" Paper presented at a symposium entitled *"IPM in Organic Systems"*, XXII International Congress of Entomology, Brisbane, Australia, 16 August 2004., available on the world wide web : http://www.organic-research.com/

Posnera Joshua L., Jon O. Baldock and Janet L. Hedtcke (2008), "Organic and Conventional Production Systems in the Wisconsin Integrated Cropping Systems Trials: I". Productivity 1990–2002, Published in *"Agron J"* 100:253-260 (2008), American Society of Agronomy, 677 S., Segoe Rd., Madison, WI 53711 USA, available on the world wide web: http://agron.scijournals.org/cgi /content/abstract/100/2/253

Prasad, R. (1999), Organic farming *vis-à-vis* modern agriculture *Curr. Sci.*, 1999, 77, 38–43.

Reganold, JP., Glover, JD., Andrews, PK ., and Hinman, HR, (2001), "Sustainability of three apple production systems" *Nature*. Vol. 410, no. 6831, pp. 926-930. 19 Apr 2001.

Singh Y. V., B. V. Singh, S. Pabbi and P. K. Singh(2007) "Impact of Organic Farming on Yield and Quality of BASMATI Rice and Soil Properties" available on world wide web http://orgprints.org/9783

Stolze Matthias, Nicolas Lampkin (2009) "Policy for organic farming: Rationale and concepts" Published by *"ELSEVIER"*. 0306-9192/$ - see front matter 2009 doi:10.1016/j.foodpol.2009.03.005 and also available on the world wide web: http://www.sciencedirect.com/science/article/B6VCB-4W2M6V2- ff57298e

Swezey Sean L.,Polly Goldman,Janet Bryer,Diego Nieto (2004), "Comparison Between Organic, Conventional, and IPM Cotton in the Northern San Joaquin Valley, California", Paper presented at a symposium entitled *"IPM in Organic Systems"*, XXII International Congress of Entomology, Brisbane, Australia, 16 August 2004 available on the world wide web: http://www.organic-research.com/

Tzouvelekas Vangelis, Christos J. Pantzios, and Christos Fotopoulos (2001), "Economic Efficiency in Organic Farming: Evidence from Cotton Farms in Viotia, Greece" *Journal of Agricultural & Applied Economics.*, Volume 33, April, 2001, Issue: 1, Pp: 35-48, available on the world wide web: http://ideas.repec.org/a/jaa/jagape/v33y2001i1p35-48.html#abstract

Wood Richard, Manfred Lenzen, Christopher Dey, Sven Lundie (2005) "A comparative study of some environmental impacts of conventional and organic farming in Australia" Published in *"ELSEVIER"*. and also available on the world wide web: http:// doi:10.1016/j.agsy.2005.09.007

Wyss E.,H. Luka,L. Pfiffner,C. Schlatter,G. Uehlinger,C. Daniel (2004), "Approaches to Pest Management in Organic Agriculture: A Case study in European Apple Orchards" Paper presented at a symposium entitled *"IPM in Organic Systems"*, XXII International Congress of Entomology, Brisbane, Australia, 16 August 2004, available on the world wide web: http://www.organic-research.com/

Chapter 3

Brief Profile of the Study Area

In this Chapter an attempt has been made to present a brief socio-economic profile of the study area.

Topography of the Select Districts

East Godavari District is located in the North Coastal part of the State of Andhra Pradesh. The District is situated on the North East of Andhra Pradesh in the geographical coordination of 16° 30′ and 18° 20′ of the Northern latitude and 81° 30′ and 82° 36 of the Eastern longitude. The District is bounded by Visakhapatnam and the state of Odisha on the North, by Bay of Bengal on the East, by West Godavari District on the South and by Khammam District on the West. The District is known as the rice bowl of Andhra Pradesh with lush paddy fields and coconut groves. It is also known as another Kerala. Its Headquarters is Kakinada. The total geographical area of the district is 10807 sq. kms.

Mahabubnagar is in Telengana part of the State and is located between 16^0 and 17^0 Northern latitude and 77^0 and 79^0 Eastern longitude. It is bounded on the North by Ranga Reddy and Nalgonda districts, on the east by Guntur district, on the South by the Krishna and the Tungabhadra rivers and on the West by Raichur and Gulbarga districts of Karnataka State. It is the second largest district in Andhra Pradesh in terms of area covered. Its Headquarters town has been named after His Excellency Mir Mahabub Ali Khan, one of the Nizams of Hyderabad State. The area of the District is 18,432 sq. kms.

Anantapur District a part of Rayalaseema lies in between 13°-40' and 15°-15' Northern latitude and 76°-50' and 78°-30' Eastern longitude. It is bounded by Bellary and Kurnool Districts on the North, Kadapa District and Kolar Districts of Karnataka on South and East respectively. The District is roughly oblong in shape, the longer side running North to South with a portion of Chitradurg District of Karnataka State intruding into it from West between Kundurpi and Amarapuram Mandals. The total geographical area of the District is 191300 sq. kms.

The other features of the study area like demographic, agro-economic, socio-economic characteristic features have been presented and analysed in the succeeding part of the Chapter.

3.1 Demographic Particulars

The demographic features of the population, viz., composition of population, sex ratio, density of population, etc. are analysed basing on

Table 3.1: Demographic Features of the Selected Districts and in Andhra Pradesh-2011

(Figures in Percentages)

S.No.	*Particulars*	*East Godavari*	*Mahabubnagar*	*Anantapur*	*Andhra Pradesh*
1.	Males	49.88	50.62	50.57	50.21
2.	Females	50.12	49.38	49.43	49.79
3.	Rural Population	76.49	89.44	74.73	72.7
4.	Urban Population	23.51	10.56	25.27	27.3
5.	SC Population	18.00	17.10	14.14	16.19
6.	ST Population	3.9	7.94	3.46	6.59
7.	Total Population	100.00 (51.51)	100.00 (40.42)	100.00 (40.83)	100.00 (846.5)
8.	Density of Population	477	219	213	308
9.	Sex Ratio	1005	975	977	992

Note: Figures in parentheses denote Population in lakhs.
Source: www.censusindia.gov.in

2011 census data. the details are presented in the Table 3.1. It can be easily found from the Table that in all the selected districts as well as in the State, the ratio of male population to female population is almost the same. It can also be observed from the Table that about 75 percent of population of the selected districts is residing in rural areas with more or less variations in the percentages, except in Anantapur, where around 90 percent of the population is residing in rural areas, resembles the rural character of the study area. As far as the density of population is concerned, East Godavari District has more density constituting 477, compared to the other districts and the State. With regard to sex ratio, which shows the availability of number of females per 1000 males, Mahabubnagar District is lagging behind (975) compared to the other districts of the State (East Godavari 1005, Anantapur 977 and Andhra Pradesh 992).

3.2 Literacy Levels

Literacy is an important variable influencing the decision making process. In the context of agriculture, a literate farmer will be more accessible to knowledge on latest developments in farm practices and there by inclined to adopt modern farming practices. In this regard, the levels of literacy of the selected districts and the State have been presented in the Table 3.2. It can be observed from the Table that the literacy levels of East Godavari District are higher (71 per cent) compared to the Anantapur (64 per cent) and Mahabubnagar (56 per cent). A close perusal of the Table reveals that East Godavari District reports very high levels of literacy rates for both males (75 per cent) and females (68 per cent), compared to other districts. Mahabubnagar and Anantapur Districts are, at the other extreme, constituting 66 per cent male literacy and 45 per cent female literacy, 74 per cent male literacy rates and 54 per cent female literacy respectively.

Table 3.2: Levels of Literacy in the Selected Districts and in Andhra Pradesh (2011 Census)

S.No.	*Persons*	*East Godavari*	*Mahabubnagar*	*Anantapur*	*Andhra Pradesh*
1.	Males	74.91	66.27	74.09	75.56
2.	Females	67.82	45.65	54.31	59.74
3.	Total	71.35	56.06	64.28	67.66

Source: As shown ante.

3.3 Occupational Pattern

The particulars of the occupational distribution of the population of the selected districts as well as the State have been presented in Table 3.3. A close observation of the Table shows that more than half of the population of the East Godavari District is unproductive (60 percent). A more or less similar picture, can be found in Anantapur District and in the State of Andhra Pradesh constituting 51 per cent and 54 percent respectively. It also reveals that in Mahabubnagar District, the percentage of unproductive population is less (47 percentage) compared to the other selected districts and the State.

Table 3.3: Occupational Distribution of the Selected Districts and in Andhra Pradesh – 2010-11

S.No.	*Particulars of Workers*	*East Godavari*	*Mahabubnagar*	*Anantapur*	*Andhra Pradesh*
1.	Total Population	(51.51) 100.00	(40.42) 100.00	(40.83) 100.00	(846.5) 100.00
2.	Percentage of Main Workers to Total Population	33.00	42.07	38.60	38.10
3.	Percentage of Marginal Workers to Total Population	6.60	11.18	10.23	7.70
4.	Percentage of Non-Workers to Total Population	60.40	46.75	51.17	54.20
5.	Percentage to Cultivators to Main Workers	14.90	15.05	27.46	27.74
6.	Percentage of Agricultural Labour to Main Workers	50.40	15.00	26.01	40.87

Note: Figures in parentheses denote Population in lakhs.
Source: Directorate of Economics and Statistics, Govt. of Andhra Pradesh, Hyderabad.

3.4 Rainfall

It is a known fact that, nature plays a dominant role in agriculture, especially in developing countries like India. Rainfall is an exogenous variable, which can neither be predicted nor be controlled. Many scientific pursuits to create artificial rains proved to be futile. Rainfall and to some extent climate, have a considerable influence on the cropping pattern, production and productivity. Adequate and timely rainfall has a positive effect on production and yield levels. Lands can be classified as arid, semi-arid and fertile on the basis of levels of rainfall. Table - 3.4 presents the data pertaining to the rainfall in the selected districts and for Andhra Pradesh from 1994-95 to 2008-09. It can be seen from the Table that the normal rain fall of the districts of Mahabubnagar and Anantapur are 604mm and 553mm respectively indicating that they receive lower rainfall compared to normal rainfall of the State (940mm). It can also be seen from the Table that East Godavari District receives 1,218 mm rain-fall, which is far higher than the State average.

Table 3.4: Average Rain-fall in the Selected Districts and in Andhra Pradesh from 1994-95 to 2008-09(Rain-fall in mm.)

Year	*East Godavari*	*Mahabubnagar*	*Anantapur*	*Andhra Pradesh*
1994-95	1546	515	377	874
1995-96	1246	762	531	971
1996-97	1616	746	750	1110
1997-98	1062	499	441	815
1998-99	1692	845	695	1128
1999-00	1012	453	521	771
2000-01	1021	668	612	925
2001-02	997	688	702	874
2002-03	707	535	290	613
2003-04	1078	624	523	936
2004-05	873	413	434	704
2005-06	1389	973	791	1147
2006-07	1168	484	408	857
2007-08	1315	845	816	1080
2008-09	1405	458	714	847
Normal Rainfall	1218	604	553	940

Source: As shown ante

3.5 Irrigation

Particulars of area irrigated under various sources in the selected districts and Andhra Pradesh have been furnished in Table-3.5. A glance at the table reveals that the major source of irrigation in East Godavari District is canals, which constitutes 49 per cent of the total operated area of the District, while in Mahabubnagar and Anantapur district, tube well / dug well, constitute 18 per cent and 8 per cent of the total operated area respectively. The State figures indicate that tube wells / dug wells irrigate about 16 per cent to total operated area, followed by canals (12 per cent).

Table 3.5: Distribution of Area Irrigated under Various Sources in the Selected Districts and in Andhra Pradesh (Tri-Annum: 2006-07 to 2008-09) (Area in Hectares)

S. No.	*Source of Irrigation*	*East Godavari*	*Mahabub nagar*	*Anantapur*	*Andhra Pradesh*
1.	Canals	2,21,588 (48.78)	28,494 (3.59)	23,524 (1.88)	16,33,873 (12.03)
2.	Tanks	14,224 (3.13)	3,976 (0.50)	5,396 (0.43)	6,11,667 (4.50)
3.	Tube Wells/Dug Wells	62,174 (13.69)	1,45,553 (18.36)	1,01,471 (8.11)	22,27,964 (16.40)
4.	Other Wells	139 (0.03)	7,797 (0.98)	12,823 (1.02)	21,453 (0.16)
5.	Lift Irrigation	3,606 (0.79)	2,103 (0.27)	45 (0.0001)	8,679 (0.06)
6.	Other Sources	891 (0.20)	5,360 (0.68)	1,577 (0.13)	1,35,457 (1.00)
7.	Net Area Irrigated	2,81,485 (66.62)	1,54,339 (19.47)	1,15,453 (11.57)	46,38,929 (34.15)
8.	Gross Area Irrigated	4,91,980	2,23,477	1,44,837	63,64,833
9.	Area Irrigated More than Once	2,11,522	50,217	29,385	17,26,066
10.	Total Operated Area	4,54,257 (100.00)	7,92,904 (100.00)	12,51,634 (100.00)	1,35,8,2000 (100.00)

Source: As shown ante

3.6 Livestock And Poultry Population

Livestock plays a vital role in the country's economy. As per 1993 State Income Estimates, the contribution of Livestock to value of output of agriculture proper (value of output of agriculture plus livestock) was 21.4 per cent at current prices. It was 25.8 per cent at all India level (State Report of Live Stock Census, 1993)[1]. The live-stock population in India is the largest in the World. In India, Andhra Pradesh occupies a prominent place with regard to livestock, which are being maintained, mainly for dairy products and for purposes of meat, hides, skins, horns, bones and wools. The particulars of live-stock and poultry population for the year 2008-09 for the selected districts and Andhra Pradesh have been presented in Table 3.7. It can be found from the Table that the percentage of buffaloes to the total live-stock population is very high in East Godavari District. In Anantapur and Mahabubnagar Districts the percentage of sheep to the total livestock population is high constituting 83 per cent and 58 per cent respectively. The State level figures show that the percentage of sheep is more constituting 43 per cent followed by buffaloes (22 per cent), cattle (19 per cent), goats(16 per cent), pigs (0.73 per cent), Other livestock (0.13 per cent), horses ponies (0.04 per cent), donkeys (0.01 per cent) and camel (0.0002 per cent)

Table 3.6: Livestock and Poultry Population of the Selected Districts and in Andhra Pradesh – 2008-09

S. No.	*Particulars*	*East Godavari*	*Mahabubnagar*	*Anantapur*	*Andhra Pradesh*
1.	Cattle	290158 (17.58)	841017 (16.82)	766455 (13.43)	11223044 (18.65)
2.	Buffaloes	975243 (59.07)	461232 (9.22)	529185 (9.27)	13271714 (22.06)
3.	Sheep	161309 (9.77)	4164497 (83.27)	3301494 (57.85)	25539452 (42.44)
4.	Goats	196446 (11.90)	685155 (13.70)	944395 (16.55)	9626012 (16.00)
5.	Horses and Ponies	49 (0.002)	5337 (0.11)	655 (0.01)	25972 (0.04)
6.	Donkeys	0 (0.00)	0 (0.00)	8588 (0.15)	8614 (0.01)
7.	Camels	1 (0.0001)	0 (0.0000)	42 (0.0007)	121 (0.0002)
8.	Pigs	27691 (1.68)	44170 (0.88)	23591 (0.41)	438653 (0.73)
9.	Other Livestock	0 (0.00)	102091 (2.04)	133020 (2.33)	75896 (0.13)
10.	Total Live Stock	1650896 (100.00)	5001250 (100.00)	5707425 (100.00)	60174771 (100.00)
11.	Total Poultry	17705685	5497731	1826856	123984716

Source: As shown ante

3.7 Land Utilisation Pattern

The analysis of land utilisation in any area is very important as it gives a wide picture of land-use pattern including the net area sown and the resultant economies contributing to the economic growth of the zone.

Particulars of land utilization pattern for the selected districts and Andhra Pradesh have been presented in Table 3.7. It is evident from the Table that the percentage of Net Area Sown to the Total Geographical Area (TGA) of East Godavari and Mahabubnagar Districts and in the State are (about 40.00 per cent) almost same, whereas in Anantapur District, it is around 57 per cent for 2008-09. As far as the forest cover of the selected districts is concerned, it is around 30 per cent of the TGA in East Godavari District, 14 per cent in Mahabubnagar, 10 per cent in Anantapur districts, and 23 per cent in the State. It can also be observed from the Table that Mahabubnagar District recorded the highest percentage of area under current fallows to its TGA, East Godavari District has recorded the highest percentage of area sown more than once and Gross Cropped Area (GCA), which can be attributed to the huge availability of water resources in the District. The State level figures show that 40 per cent of the TGA is net area sown followed by forest lands (23 per cent), land put to non-agricultural uses (9.64 per cent), current fallows (9.54 percent), barren and uncultivable land (7.47 per cent), other fallow lands (5.41 per cent) etc.

Table 3.7: Land Utilisations in the Selected Districts and in Andhra Pradesh 2008-09 (Figures in Percentages)

S. No.	*Category*	*East Godavari*	*Mahabu-bnagar*	*Anantapur*	*Andhra Pradesh*
1.	Forest	29.91	13.87	10.30	22.58
2.	Barren and Uncultivable Land	7.32	4.80	9.68	7.47
3.	Land put to Non-Agricultural Uses	12.47	4.41	6.31	9.64
4.	Cultivable Waste	1.67	1.04	2.76	2.36
5.	Permanent Pastures and other Grazing Lands	1.91	0.95	0.47	2.07
6.	Miscellaneous Tree Crops and Groves not Included in Net Area Sown	0.80	0.35	0.49	1.09
7.	Current Fallows	2.16	27.25	8.76	9.54
8.	Other Fallows	3.13	7.14	4.63	5.41
9.	Net Area Sown	40.63	40.18	56.61	39.84
10.	Total Geographical Area	(437530) 100.00	(746234) 100.00	(774494) 100.00	(11135425) 100.00
11.	Area Sown More than Once	32.16	3.74	3.70	10.77
12.	Gross Cropped Area	72.79	43.93	60.31	50.28

Note: Figures in parentheses denote Ha. of Land.
Source: As Shown ante

3.8 Area Under Principal Crops

The particulars of area under principal crops in the selected districts and Andhra Pradesh as an average of three years of period i.e., 2006-07 to 2008-09 have been shown in Table 3.8. It is evident from the table that the area under the main staple food stuff of Andhras i.e., paddy is around Ha. 3.9 lakh in East Godavari District, Ha. 1.6 lakh in Mahabubnagar and Ha. 0.41 lakh in Anantapur districts which constitutes around 10 per cent, one percent and four percent of the area under paddy in Andhra Pradesh during 2006-07 to 2008-09. It is also evident from the Table that around 50 per cent of the total area under groundnut in Andhra Pradesh is sown in Anantapur District and 25 per cent of the total area is under Jowar, 16 per cent of the total area under redgram and 15 per cent of the area under maize in Andhra Pradesh are sown in Mahabubnagar District, which shows dependence of the people in selected districts on various principal crops.

Table 3.8: Area under Principal Crops in the Selected Districts and in Andhra Pradesh(Tri-Annum: 2006-07 to 2008-09)(Area in Hectares)

S. No	*Name of the Crop*	*East Godavari*	*Mahabubnagar*	*Anantapur*	*Andhra Pradesh*
1.	Rice	397680 (9.66)	169667 (4.12)	41667 (1.01)	4116350 (29.76)
2.	Jowar	730 (0.21)	85000 (24.70)	24667 (7.17)	344088 (2.48)
3.	Bajra	1550 (2.40)	0 (0.00)	0 (0.00)	64718 (0.46)
4.	Maize	4840 (0.61)	123667 (15.70)	5667 (0.72)	787870 (5.69)
5.	Total Cereals	421279 (53.55)	333764 (41.22)	75545 (6.54)	5671109 (41.00)
6.	Redgram	897 (0.21)	71333 (16.37)	32333 (7.42)	435722 (3.15)
7.	Bengal Gram	0 (0.00)	0 (0.00)	77333 (12.62)	612844 (4.43)
8.	Total Pulses	79289 (10.07)	125146 (15.45)	109199 (9.46)	1771473 (12.80)
9.	Groundnut	687 (0.04)	120000 (7.35)	810000 (49.63)	1631964 (11.80)
10.	Total Oil Seeds	58187 (7.39)	215218 (26.58)	914737 (79.28)	2728226 (19.72)
11.	Chillies	1373 (0.64)	0 (0.00)	2000 (0.94)	213357 (1.54)
12.	Sun Flower	0 (0.00)	0 (0.00)	49667 (11.55)	429966 (3.10)

Source: As shown ante

3.9 Cropping Pattern

Cropping pattern means the proportion of area under different crops at a point of time. Any change in cropping pattern implies a change in the proportion of area under different crops. Cropping pattern of any region depends on physical characteristics such as soil, climate, weather, rainfall etc. Apart from soil and climatic conditions, the cropping pattern of a region will also depend on the nature and irrigation facilities, available locally.

Table 3.9: Area under Food and Non- Food Crops in the Selected Districts and in Andhra Pradesh(Tri-Annum: 2006-07 to 2008-09)(Area in Hectares)

S.No	*Name of the Crop*	*East Godavari*	*Mahabubnagar*	*Anantapur*	*Andhra Pradesh*
1.	Paddy	409200 (52.02)	144760 (17.88)	48680 (4.22)	4386900 (31.72)
2.	Jowar	630 (0.08)	50440 (6.23)	21110 (1.83)	4898620 (35.42)
3.	Maize	0 (0.00)	125980 (15.56)	0 (0.00)	851930 (6.16)
4.	Bajra	1170 (0.15)	0 (0.00)	0 (0.00)	1040020 (7.52)
5.	Redgram	780 (0.10)	80000 (9.88)	34030 (2.95)	442560 (3.20)
6.	Bengal Gram	0 (0.00)	0 (0.00)	73030 (6.33)	607140 (4.39)
7.	Groundnut	550 (0.07)	99500 (12.29)	890.50 (75.45)	1766100 (12.77)
8.	Total Food Crops	686790 (87.31)	499960 (61.75)	231670 (20.08)	9122320 (65.96)
9.	Total Non - Food Crops	99820 (12.69)	309680 (38.25)	922060 (79.92)	4707760 (34.04)
10.	Total Food and Non-Food Crops	786610 (100.00)	809640 (100.00)	1153730 (100.00)	13830080 (100.00)

Source: As shown ante

Table 3.9 depicts the area under food and non-food crops in the selected districts and in the State for 2008-09. It is apparent from the Table that area under food crops is around 87 per cent of the total cropped area in East Godavari District, which is more than the State average (around 66 per cent). It is 62 per cent in Mahabubnagar and 20 per cent in Anantapur districts. Another interesting thing that can be observed from the Table is that the percentage of area under non-food crops is around 80 per cent to the total cropped area in Anantapur District. A crop wise analysis shows that 52 percentage of total cropped area is under paddy cultivation in East Godavari District, whereas it is just four and 18 per cent in Mahabubnagar and Anantapur districts respectively. In Anantapur District, around 75 per cent of total cropped area is under Groundnut cultivation, whereas it is just 0.01 per cent in East Godavari and 12 per cent in Mahabubnagar districts.

3.10 Land Holding Particulars

The particulars of area of operational holdings in the selected districts and in Andhra Pradesh for the year 2008-09 have been presented in Table 3.10. It is apparent from the Table that the average size of the agricultural holding in East Godavari District is Ha. 0.76,which is lower than the average size of the land holdings in Anantapur Ha. 1.93 and Mahabubnagar Ha. 1.55 as well as the State average (Ha.1.20). Another interesting point that can be observed from the Table is that the percentage of marginal land holdings to total land-holdings is the highest, in East Godavari District (38 per cent) followed by small land holdings (24 per cent), semi-medium (21 per cent), medium (13 per cent) and large (3.5 per cent), whereas in Mahabubnagar and Anantapur districts, the percentage of semi-medium land holdings is higher constituting 29 per cent and 32 per cent respectively. A more or less similar picture with slight variations in percentages can be found at the State level also.

Table 3.10: Area of Operational Holdings in the Selected Districts and in Andhra Pradesh – 2008-09(Area in Hectares)

S.No.	*Category*	*East Godavari*	*Mahabu-bnagar*	*Anantapur*	*Andhra Pradesh*
1.	Marginal (Up to 1.0)	198892 (38.30)	191070 (16.03)	127147 (10.01)	3287034 (27.69)
2.	Small (1.0 - 2.0)	125188 (24.11)	288213 (24.18)	305773 (24.07)	3730303 (31.43)
3.	Semi-Medium (2.0 - 4.0)	108444 (20.88)	340074 (28.53)	411342 (32.38)	3835072 (32.31)
4.	Medium (4.0 - 10.0)	68283 (13.15)	267756 (22.46)	307519 (24.21)	2758745 (23.24)
5.	Large (10.0 and Above)	18459 (3.55)	104955 (8.80)	118447 (9.32)	877734 (7.39)
Total		519255 (100.00)	1192068 (100.00)	1270228 (100.00)	11869949 (100.00)
Average Size of Holding		0.76	1.55	1.93	1.20

Source: As shown ante

Summary

- The literacy levels of East Godavari District are higher for both males and females compared to Anantapur and Mahabubnagar.
- While more than one half of the population of Andhra Pradesh, East Godavari and Anantapur Districts is unproductive, it is lower in Mahabubnagar.
- The major source of irrigation in East Godavari District is canals, which constitutes 49 per cent of the total operated area of the District, while in Mahabubnagar and Anantapur district, tube well / dug well, constitute 18 per cent and 8 per cent of the total operated area respectively. The State figures indicate that tube wells / dug wells irrigate about 16 per cent to total operated area, followed by canals (12 per cent).
- The percentage of buffaloes in the total live-stock population is very high in East Godavari District, while in Anantapur and Mahabubnagar districts, the percentage of sheep to the total livestock population constitutes 83 per cent and 58 per cent respectively.

The above analysis of the socio-economic profile of the study area reveals that the conditions prevailed in East Godavari District like literacy rate, percentage of the aged and experienced population to total population, average rain-fall, irrigation facilities and availability of dung (organic manure), are more favorable for organic farming compared to the other selected districts. Thus, it can be concluded that East Godavari District is congenial for organic farming when compared to the other two selected districts. So, it can be hypothesized that the organic farmers in East Godavari District are in an advantageous position in relation to efficient input-use compared to other farmers in Mahabubnagar and Anantapur.

Note

Report of Live Stock Census 1993, Published by Directorate of Economics and Statistics, Govt. of A.P.

Chapter 4

Profile of the Sample Households

For any research in social sciences, it is mandatory to analyse the socio-economic characteristic features of sample households like age, education, farm size, assets, experience in farming practice, liabilities etc. to have a clear idea on the economy and to come to any reasonable conclusions. Hence, in this Chapter an attempt has been made to analyse the socio-economic features of the sample households.

Distribution of sample households by farming practice has been presented in Table 4.1.

Table 4.1: Distribution of Sample Households by Farming Practice and Crops Grown

S. No.	*Name of the Crop*	*Organic*	*Conventional*	*Total*
1.	Paddy	150 (60.00)	100 (40.00)	250 (100.00)
2.	Redgram	100 (66.67)	50 (33.33)	150 (100.00)
3.	Groundnut	100 (66.67)	50 (33.33)	150 (100.00)
	Total	350 (63.64)	200 (36.36)	550 (100.00)

Note: Figures in parentheses indicate percentages to total.

Distribution of sample households by crop wise, farm size wise and farming practice wise has been presented in Table 4.2. It can be observed from the Table that out of the total 350 selected organic farming households,

128 households are small farmers, 141 households are medium farmers and the remaining 81 households are large farmers constituting 37, 40 and 23 percent respectively. Out of the total 200 selected conventional farming households, 34 per cent are small farming households, 42 per cent are medium farming households and the remaining are large farming households. A crop-wise and farm-size wise analysis shows that around 37 percent of the total organic paddy growing farming households are small farmers, 44 percent are medium farmers and the remaining 19 per cent are large farmers. With regard to the conventional paddy growing farming households, a more or less similar picture can be found. As far as the groundnut growing farmers is concerned, there is not much of a difference between the two groups of farmers by farm-size. With regard to the redgram growing farming households, a more or less similar picture can be found in between the two groups of farming households. It can be concluded from the above analysis that, there is not much of a difference, in proportion between the organic and conventional farming households with regard to the distribution of households by farm size and crop wise.

Table 4.2: Distribution of Sample Households by Crop, Farm size and Farming Practice.

S. No.	*Crop*	*Small*	*Medium*	*Large*	*All Farms*
Organic Farmers					
1.	Paddy	55 (36.67)	66 (44.00)	29 (19.33)	150 (100.00)
2.	Redgram	38 (38.00)	34 (34.00)	28 (28.00)	100 (100.00)
3.	Groundnut	35 (35.00)	41 (41.00)	24 (24.00)	100 (100.00)
	Total	128 (36.57)	141 (40.29)	81 (23.14)	350 (100.00)
Conventional Farmers					
1.	Paddy	39 (39.00)	36 (36.00)	25 (25.00)	100 (100.00)
2.	Redgram	14 (28.00)	25 (50.00)	11 (22.00)	50 (100.00)
3.	Groundnut	16 (32.00)	22 (44.00)	12 (24.00)	50 (100.00)
	Total	69 (34.50)	83 (41.50)	48 (24.00)	200 (100.00)

Note: Figures in parentheses indicate percentages to totals.

The structure of land holdings for paddy, groundnut and redgram farmers by farm size and farming practice-wise has been presented in Tables 4.3.1 to 4.3.3. It can be easily traced out from these tables that for all the selected households from organic category the total operated area includes owned lands only. No pieces of land are either leased-in or leased-out lands for the reason that these farmers are adopting organic farming only on their own fields as the owners of the leased-in lands may not allow them to experiment on their fields. But in the conventional category the situation is different, when pieces of land are leased-in for cultivation purposes.

Another important point that can be observed from these three tables is that no selected household has leased-out its land in both the categories. In the organic category the total operated area under paddy, redgram and groundnut are Ac. 1190.5, Ac. 815.5 and Ac. 896.7 respectively, whereas in the conventional category, the figures are Ac. 822.1, Ac. 403.7 and Ac. 419.9 for paddy, redgram and groundnut respectively.

Table 4.3.1: Structure of Land Holdings - Selected Paddy Farmers(Area in Acres)

Organic Farmers					
S. No.	*Item*	*Small*	*Medium*	*Large*	*All Farms*
1.	No. of House Holds	55	66	29	150
2.	Owned Land	179.8	492.6	518.1	1190.5
3.	Leased - In	0	0	0	0
4.	Leased - Out	0	0	0	0
5.	Operated Area	179.8	492.6	518.1	1190.5
6.	Average Size of Holding	3.27	7.46	17.87	7.94
Conventional Farmers					
1.	No. of House Holds	39	36	25	100
2.	Owned Land	121.6	206.5	370.3	698.4
3.	Leased - In	12.3	70.3	41.1	123.7
4.	Leased - Out	0	0	0	0
5.	Operated Area	133.9	276.8	411.4	822.1
6.	Average Size of Holding	3.43	7.69	16.46	8.22

Note: Figures in parentheses indicate percentages to totals.

Table 4.3.2: Structure of Land Holdings - Selected Redgram Farmers (Area in Acres)

S. No.	*Item*	*Small*	*Medium*	*Large*	*All Farms*
Organic Farmers					
1.	No. of House Holds	38	34	28	100
2.	Owned Land	117.9	247.8	450.2	815.9
3.	Leased - In	0	0	0	0
4.	Leased - Out	0	0	0	0
5.	Operated Area	117.9	247.8	450.2	815.9
6.	Average Size of Holdings	3.10	7.29	16.08	8.16
Conventional Farmers					
1.	No. of House Holds	14	25	11	50
2.	Owned Land	41.2	143.5	153.4	338.1
3	Leased -In	8.2	45.1	12.3	65.6
4.	Leased Out	0	0	0	0
5.	Operated Area	49.4	188.6	165.7	403.7
6.	Average Size of Holdings	3.53	7.54	15.06	8.07

Farm size-wise analysis of these three tables has revealed that only 15 percent of the total operated area under paddy is being cultivated by the small farmers, whereas the remaining 85 per cent is being cultivated by medium (41 per cent) and large (44 percent) farmers in organic category for paddy. With regard to redgram and groundnut, a more or less similar picture can be found. As far as the farm size wise analysis of the three tables for conventional category is concerned, it can be found that only 16, 12 and 11 per cent of the total paddy operated area is sown by the small farmers for paddy, redgram and groundnut respectively, whereas the remaining operated area is sown by hands of medium and large farmers for all the three crops respectively.

With regard to the average size of the land holdings, it is Ac. 3.27 for small farmers, Ac. 7.46 for medium farmers and Ac. 17.87 for large farmers for paddy in organic farmers' category, while it is Ac. 3.43 for small farmers, Ac. 7.69 for medium farmers and Ac. 16.46 for large farmers in conventional farmers' category. For other crops, a more or less similar picture, can be found.

Table 4.3.3: Structure of Land Holdings - Selected Groundnut Farmers
(Area in Acres)

Organic Farmers					
S. No.	*Item*	*Small*	*Medium*	*Large*	*All Farms*
1.	No. of House Holds	35	41	24	100
2.	Owned Land	116.5	306.2	474.0	896.7
3.	Leased -In	0	0	0	0
4.	Leased Out	0	0	0	0
5.	Operated Area	116.5	306.2	474.0	896.7
6.	Average Size of Holdings	3.33	7.47	19.75	8.97
Conventional Farmers					
S. No.	*Item*	*Small*	*Medium*	*Large*	*All Farms*
1.	No. of House Holds	16	22	12	50
2.	Owned Land	40.6	129.5	150.2	320.3
3.	Leased -In	6.3	37.7	55.6	99.6
4.	Leased Out	0	0	0	0
5.	Operated Area	46.9	167.2	205.8	419.9
6.	Average Size of Holdings	2.93	7.6	17.15	8.4

The distribution of family members of the sample households by crop wise and farming practice wise has been presented in Table 4.4. A glance at Table 4.4.1 has revealed that, out of the total 635 family members of the organic farmers category of paddy farmers, around 46 per cent of the members are males, 38 per cent are females and the remaining 16 per cent are children in the group of below five years. A close perusal at the Table by farm size also revealed a more or less same picture. With regard to the conventional farmers category of paddy farmers, out of the total 467 members, 44 per cent are males, 36 per cent are females and the remaining 20 per cent are children in the age group of below five years.

A glance at Table 4.4.2 depicts that, out of the total 458 family members of the organic farming category of redgram farmers, around 50 per cent of the members are males, 41 per cent are females and the remaining nine per cent are children with age group of below five years. A close perusal at the Table by farm size also revealed a more or less same picture.

With regard to the conventional farming category of redgram farmers, out of the total family members 249, 51 per cent are males (out of which 28 per cent belong to small farmers category, 51 per cent belong to medium farming category and the remaining 21 per cent are belong to large farming category), 35 per cent are females(out of which 27 per cent belong to small farmers category, 54 per cent belongs to medium farming category and the remaining 19 per cent belong to large farming category) and the remaining 14 per cent are children in the age group of below five years(out of which 33 per cent belong to small farmers category, 44 per cent belongs to medium farming category and the remaining 55 per cent belong to large farmers category).

Table 4.4.3 reveals that, out of the total 428 family members of the organic farming category of groundnut farmers, around 50 per cent of them are males, 40 per cent are females and the remaining 10 per cent are children in the age group of below five years. A close perusal at the Table by farm size also revealed a more or less same picture. With regard to the conventional farmers' category of groundnut farmers, out of the total 238 members, 47 per cent are males, 39 per cent are females and the remaining 14 per cent are children with in the age group of below five years.

Table 4.5 furnish particulars of distribution of sample farmers by age group, farm size and farming practice category. It can be easily traced out from the Table - 4.5.1 that out of the total 150 organic paddy farmers, 30 per cent of them are in the age group of below 30 years, 43 per cent in the age group of 30-55 years and the remaining 27 per cent are of above 55 years. A more or less similar picture can also be found with regard to the conventional paddy farmers also.

It can be found from the Table 4.5.2 that out of the total 100 organic redgram growing farmers, 36 per cent of them are in the age group of below 30 years, 44 per cent are in the age group of 30-55 years and the remaining 20 per cent are in the age group of above 55 years. With regard to the conventional redgram farmer category, 28 per cent of them are in the age group of below 30 years, 36 per cent are in the age group of 30-55 years and the remaining 36 per cent are in the age group of above 55 years.

Table 4.4.1: Distribution of Family Members –Selected Paddy Farmers

Organic Farmers					
S. No.	*Sex*	*Small*	*Medium*	*Large*	*All Farms*
1.	Males	94 (44.98)	130 (46.76)	68 (45.95)	292 (45.98)
2.	Females	80 (38.28)	108 (38.85)	54 (36.49)	242 (38.11)
3.	Children (0-5 Years)	35 (16.75)	40 (14.39)	26 (17.57)	101 (15.91)
	Total	209 (100.00)	278 (100.00)	148 (100.00)	635 (100.00)
Conventional Farmers					
1.	Males	74 (44.05)	88 (43.78)	44 (44.90)	206 (44.11)
2.	Females	62 (36.90)	72 (35.82)	36 (36.73)	170 (36.40)
3.	Children (0 - 5 Years)	32 (19.05)	41 (20.40)	18 (18.37)	91 (19.49)
	Total	168 (100.00)	201 (100.00)	98 (100.00)	467 (100.00)

Note: Figures in parentheses indicate percentages to total.

Table 4.4.2: Distribution of Family Members – Selected Redgram Farmers

Organic Farmers					
S. No.	*Sex*	*Small*	*Medium*	*Large*	*All Farms*
1.	Males	63 (45.99)	84 (53.16)	83 (50.92)	230 (50.22)
2.	Females	59 (43.07)	60 (37.97)	68 41.72	187 (40.83)
3.	Children (0-5 Years)	15 (10.95)	14 (8.86)	12 7.36	41 (8.95)
	Total	137 (100.00)	158 (100.00)	163 100.00	458 (100.00)
Conventional Farmers					
1.	Males	38 (55.07)	71 (51.35)	29 (58.00)	138 (51.35)
2.	Females	28 (40.58)	55 (35.14)	19 (38.00)	102 (35.14)
3.	Children (0 - 5 Years)	3 (4.35)	4 (13.51)	2 (4.00)	9 (13.51)
	Total	69 (100.00)	130 (100.00)	50 (100.00)	249 (100.00)

Note: Figures in parentheses indicate percentages to total.

Table 4.4.3: Distribution of Family Members – Selected Groundnut Farmers

S. No.	Item	Small	Medium	Large	All Farms
Organic Farmers					
1.	Males	72 (50.00)	79 (50.64)	62 (48.44)	213 (49.77)
2.	Females	56 (38.89)	64 (41.03)	49 (38.28)	169 (39.49)
3.	Children (0-5 Years)	16 (11.11)	13 (8.33)	17 (13.28)	46 (10.75)
	Total	144 (100.00)	156 (100.00)	128 (100.00)	428 (100.00)
Conventional Farmers					
1.	Males	38 (51.35)	53 (44.54)	22 (48.89)	113 (47.48)
2.	Females	26 (35.14)	48 (40.34)	18 (40.00)	92 (38.66)
3.	Children (0 - 5 Years)	10 (13.51)	18 (15.13)	5 (11.11)	33 (13.87)
	Total	74 (100.00)	119 (100.00)	45 (100.00)	238 (100.00)

Note: Figures in parentheses indicate percentages to total.

Table 4.5.1: Distribution of the Head of the Sample Households – Paddy Farmers (Age wise)

S. No.	Item	Small	Medium	Large	All Farms
Organic Farmers					
1.	Up to 30 Years	16 (29.09)	21 (31.82)	8 (27.59)	45 (30.00)
2.	30 – 55 Years	22 (40.00)	31 (46.97)	11 (37.93)	64 (42.67)
3.	Above 55 Years	17 (30.91)	14 (21.21)	10 (34.48)	41 (27.33)
	Total	55 (100.00)	66 (100.00)	29 (100.00)	150 (100.00)
Conventional Farmers					
1	Up to 30 Years	14 (35.90)	7 (19.44)	9 (36.00)	30 (30.00)
2.	30 – 55 Years	19 (48.72)	21 (58.33)	6 (24.00)	46 (46.00)
3.	Above 55 Years	6 (15.38)	8 (22.22)	10 (40.00)	24 (24.00)

Note: Figures in parentheses indicate percentages to total.

Table 4.5.2: Distribution of the Head of the Sample Households – Redgram Farmers (Age wise)

S. No.	Item	Small	Medium	Large	All Farms
Organic Farmers					
1.	Up to 30 Years	15 (39.47)	9 (26.47)	12 (42.86)	36 (36.00)
2.	30 – 55 Years	18 (47.37)	17 (50.00)	9 (32.14)	44 (44.00)
3.	Above 55 Years	5 (13.16)	8 (23.53)	7 (25.00)	20 (20.00)
	Total	38 (100.00)	34 (100.00)	28 (100.00)	100 (100.00)
Conventional Farmers					
1.	Up to 30 Years	4 (28.57)	6 (24.00)	4 (36.36)	14 (28.00)
2.	30 – 55 Years	3 (21.43)	9 (36.00)	6 (54.55)	18 (36.00)
3.	Above 55 Years	7 (50.00)	10 (40.00)	1 (9.09)	18 (36.00)
	Total	14 (100.00)	25 (100.00)	11 (100.00)	50 (100.00)

Note: Figures in parentheses indicate percentages to total.

It can be found from the Table 4.5.3 that out of the total 100 organic groundnut farmers, 26 per cent of them are in the age group of below 30 years, 45 per cent are in the age group of 30-55 years and the remaining 29 per cent are in the age group of above 55 years. A more or less similar picture can be found with regard to the conventional groundnut farmer category also.

Table 4.6 depicts the particulars of literacy levels of the heads of the sample households according to farming practice and farm size. The level of literacy of farmers gives one a picture on the rationality of the farmers on various issues of farming practice. It can be found that in both organic and conventional category, no farmer is illiterate and they are literates at least with primary education.

Table 4.5.3: Distribution of the Head of the Sample Households – Groundnut Farmers (Age wise)

S. No.	Item	Small	Medium	Large	All Farms
Organic Farmers					
1.	Up to 30 Years	7 (20.00)	13 (31.71)	6 (25.00)	26 (26.00)
2.	30 – 55 Years	14 (40.00)	19 (46.34)	12 (50.00)	45 (45.00)
3.	Above 55 Years	14 (40.00)	9 (21.95)	6 (25.00)	29 (29.00)
	Total	35 (100.00)	41 (100.00)	24 (100.00)	100 (100.00)
Conventional Farmers					
1.	Up to 30 Years	4 (25.00)	6 (27.27)	4 (33.33)	14 (28.00)
2.	30 – 55 Years	8 (50.00)	10 (45.45)	6 (50.00)	24 (48.00)
3.	Above 55 Years	4 (25.00)	6 (27.27)	2 (16.67)	12 (24.00)
	Total	16 (100.00)	22 (100.00)	12 (100.00)	50 (100.00)

Note: Figures in parentheses indicate percentages to total.

A glance at all these tables has revealed that out of 150 organic paddy farmers (Table 4.6.1), 30 per cent of them had primary level education, 53 per cent had secondary level education, while 14 per cent had higher level education and the remaining three percent had technical education. In the conventional paddy farmers category, 43 per cent of them had primary education, 42 per cent had secondary education, 11 per cent had higher education and four per cent had technical education, indicating that there exists homogeneity in between in the organic and conventional categories of farmers in the study area.

As far as the literacy levels of the organic redgram farming households are concerned (Table 4.6.2), it can be observed that out of 100 farmers, 38 per cent had primary education, 41 per cent had secondary level education, 14 per cent had higher level of education and seven per cent of them had technical education. Farm size wise analysis has also revealed a more or less similar picture. A similar pattern could also be discernable conventional redgram farmers' category.

Table 4.6.1: Distribution of Literacy Levels of Heads of the Sample Households – Paddy Farmers

S. No.	Item	Small	Medium	Large	All Farms
Organic Farmers					
1.	Illiterates	0 (0.00)	0 (0.00)	0 (0.00)	0 (0.00)
2.	Primary	14 (25.45)	21 (31.82)	10 (34.48)	45 (30.00)
3.	Secondary	31 (56.36)	39 (59.09)	9 (31.03)	79 (52.67)
4.	Higher	9 (16.36)	5 (7.58)	7 (24.14)	21 (14.00)
5.	Technical	1 (1.82)	1 (1.52)	3 (10.34)	5 (3.33)
	Total	55 (100.00)	66 (100.00)	29 (100.00)	150 (100.00)
Conventional Farmers					
1.	Illiterates	0 (0.00)	0 (0.00)	0 (0.00)	0 (0.00)
2.	Primary	14 (35.90)	20 (55.56)	9 (36.00)	43 (43.00)
3.	Secondary	19 (48.72)	11 (30.56)	12 (48.00)	42 (42.00)
4.	Higher	5 (12.82)	4 (11.11)	2 (8.00)	11 (11.00)
5.	Technical	1 (2.56)	1 (2.78)	2 (8.00)	4 (4.00)
	Total	39 (100.00)	36 (100.00)	25 (100.00)	100 (100.00)

Note: Figures in brackets indicate percentages to total.

As far as the literacy levels of the organic groundnut farming households is concerned (Table 4.6.3), it can be noted that out of 100 organic groundnut farmers, 35 per cent had primary education, 50 per cent had secondary level education, 11 per cent had higher level of education and a very negligible percentage of them had technical education. Farm size wise analysis also revealed a more or less similar picture. With regard to the literacy levels of the conventional groundnut farmers, out of 50 farmers, 44 per cent had primary level education, 38 per cent had secondary level education, 10 per cent had higher level education and the remaining eight per cent had technical education.

Table 4.6.2: Distribution of Literacy Levels of Heads of the Sample Households – Redgram Farmers

S. No.	Item	Small	Medium	Large	All Farms
Organic Farmers					
1.	Illiterates	0 (0.00)	0 (0.00)	0 (0.00)	0 (0.00)
2.	Primary	13 (34.21)	14 (41.18)	11 (39.29)	38 (38.00)
3.	Secondary	16 (42.11)	15 (44.12)	10 (35.71)	41 (41.00)
4.	Higher	6 (15.79)	3 (8.82)	5 (17.86)	14 (14.00)
5.	Technical	3 (7.89)	2 (5.88)	2 (7.14)	7 (7.00)
	Total	38 (100.00)	34 (100.00)	28 (100.00)	100 (100.00)
Conventional Farmers					
1.	Illiterates	0 (0.00)	0 (0.00)	0 (0.00)	0 (0.00)
2.	Primary	7 (50.00)	8 (32.00)	4 (36.36)	19 (38.00)
3.	Secondary	4 (28.57)	11 (44.00)	5 (45.45)	20 (40.00)
4.	Higher	1 (7.14)	5 (20.00)	1 (9.09)	7 (14.00)
5.	Technical	2 (14.29)	1 (4.00)	1 (9.09)	4 (8.00)
	Total	14 (100.00)	25 (100.00)	11 (100.00)	50 (100.00)

Note: Figures in parentheses indicate percentages to total.

Levels of literacy of the family members of selected households may also have some impact on the decision making in farm management. So, literacy levels of family members of the sample households according to crop-wise are presented in Table 4.7. It can be seen from the Table 4.7.1 that out of 635 family members of the selected organic paddy growing farmers, a negligible percentage (only 0.94 per cent) of population are illiterates. It can also be seen from the Table that secondary level education occupied a major share in the total population with a percentage of 52 per cent, followed by primary level education (21 per cent), higher level of education (19 percent) and technical education (4 per cent). A more or less similar picture can also be traced out from the farm size wise analysis also. With regard to the family members of the conventional paddy farmers, 18 percent of them had primary level of education, 47 percent had secondary level education, 26 percent had higher level of education and eight percent had technical education. A more or less similar picture can be traced out from the farm size wise analysis also.

Table 4.6.3: Distribution of Literacy Levels of Heads of the Sample Households – Groundnut Farmers

Organic Farmers					
S. No.	*Item*	*Small*	*Medium*	*Large*	*All Farms*
1.	Illiterates	0 (0.00)	0 (0.00)	0 (0.00)	0 (0.00)
2.	Primary	12 (34.29)	15 (36.59)	8 (33.33)	35 (35.00)
3.	Secondary	16 (45.71)	22 (53.66)	12 (50.00)	50 (50.00)
4.	Higher	5 (14.29)	3 (7.32)	3 (12.50)	11 (11.00)
5.	Technical	2 (5.71)	1 (2.44)	1 (4.17)	4 (4.00)
6.	Total	35 (100.00)	41 (100.00)	24 (100.00)	100 (100.00)
6.	Total	35 (100.00)	41 (100.00)	24 (100.00)	100 (100.00)
Conventional Farmers					
1.	Illiterates	0 (0.00)	0 (0.00)	0 (0.00)	0 (0.00)
2.	Primary	5 (31.25)	12 (54.55)	5 (41.67)	22 (44.00)
3.	Secondary	7 (43.75)	8 (36.36)	4 (33.33)	19 (38.00)
4.	Higher	2 (12.50)	1 (4.55	2 (16.67)	5 (10.00)
5.	Technical	2 (12.50)	1 (4.55)	1 (8.33)	4 (8.00)
	Total	16 (100.00)	22 (100.00)	12 (100.00)	50 (100.00)

Note: Figures in parentheses indicate percentages to total.

It can be seen from the Table-4.7.2 that out of 458 family members of the selected households for organic redgram growing farmers, 27 percent of them had primary level of education, 48 percent had secondary level education, 20 percent had higher level of education and only four percent had technical education. A more or less similar picture can be traced out from the farm size wise analysis also. With regard to the conventional

Table 4.7.1: Distribution of Literacy Levels of Family Members - Sample Households (Paddy Farmers)

Organic Farmers					
S. No.	*Item*	*Small*	*Medium*	*Large*	*All Farms*
1.	Illiterates	3 (1.44)	2 (0.72)	1 (0.68)	6 (0.94)
2.	Primary	43 (20.57)	56 (20.14)	32 (21.62)	131 (20.63)
3.	Secondary	112 (53.59)	145 (52.16)	71 (47.97)	328 (51.65)
4.	Higher	40 (19.14)	51 (18.35)	32 (21.62)	123 (19.37)
5.	Technical	11 (5.26)	24 (8.63)	12 (8.11)	47 (7.40)
	Total	209 (100.00)	278 (100.00)	148 (100.00)	635 (100.00)
Conventional Farmers					
1.	Illiterates	1 (0.60)	2 (1.00)	1 (1.02)	4 (0.86)
2.	Primary	28 (16.67)	39 (19.40)	19 (19.39)	86 (18.42)
3.	Secondary	92 (54.76)	102 (50.75)	24 (24.49)	218 (46.68)
4.	Higher	35 (20.83)	39 (19.40)	48 (48.98)	122 (26.12)
5.	Technical	12 (7.14)	19 (9.45)	6 (6.12)	37 (7.92)
	Total	168 (100.00)	201 (100.00)	98 (100.00)	467 (100.00)

Note: Figures in parentheses indicate percentages to total.

redgram farmers, out of the 249 members, 24 percent of them had primary level of education, 57 percent had secondary level education, 12 percent had higher level of education and only five percent had technical education. Only two percent of them are illiterates, which is a negligible share. A more or less similar picture can be traced out from the farm size wise analysis also.

Table 4.7.2: Distribution of Literacy Levels of Family Members - Sample Households(Redgram Farmers)

Organic Farmers					
S. No.	*Item*	*Small*	*Medium*	*Large*	*All Farms*
1.	Illiterates	1 (0.73)	1 (0.63)	2 (1.23)	4 (0.87)
2.	Primary	48 (35.04)	35 (22.15)	41 (25.15)	124 (27.07)
3.	Secondary	61 (44.53)	79 (50.00)	81 (49.69)	221 (48.25)
4.	Higher	20 (14.60)	34 (21.52)	36 (22.09)	90 (19.65)
5.	Technical	7 (5.11)	9 (5.70)	3 (1.84)	19 (4.15)
	Total	137 (100.00)	158 (100.00)	163 (100.00)	458 (100.00)
Conventional Farmers					
1.	Illiterates	2 (2.90)	2 (1.54)	2 (4.00)	6 (2.41)
2.	Primary	21 (30.43)	27 (20.77)	11 (22.00)	59 (23.69)
3.	Secondary	35 (50.72)	79 (60.77)	28 (56.00)	142 (57.03)
4.	Higher	9 (13.04)	14 (10.77)	6 (12.00)	29 (11.65)
5.	Technical	2 (2.90)	8 (6.15)	3 (6.00)	13 (5.22)
	Total	69 (100.00)	130 (100.00)	50 (100.00)	249 (100.00)

Note: Figures in parentheses indicate percentages to total.

It can be seen from the table-4.7.3 that out of 428 family members of the selected households for organic groundnut growing farmers, 25 percent of them had primary level of education, 45 percent had secondary level education, 21 percent had higher level of education and only eight percent had technical education. A more or less similar picture can also be traced out from the farm size wise analysis also. With regard to the conventional groundnut farmers, 25 percent of them had primary level of education, 51 percent had secondary level education, 17 percent had higher level of education and only five percent had technical education. Only two percent of them are illiterates, which is a negligible share. A more or less similar picture can be traced out from the farm size wise analysis also

Table 4.7.3: Distribution of Literacy Levels of Family Members - Sample Households (Groundnut Farmers)

S. No.	*Item*	*Small*	*Medium*	*Large*	*All Farms*
Organic Farmers					
1.	Illiterates	1 (0.69)	2 (1.28)	1 (0.78)	4 (0.93)
2.	Primary	35 (24.31)	49 (31.41)	22 (17.19)	106 (24.77)
3.	Secondary	59 (40.97)	72 (46.15)	63 (49.22)	194 (45.33)
4.	Higher	41 (28.47)	23 (14.74)	26 (20.31)	90 (21.03)
5.	Technical	8 (5.56)	10 (6.41)	16 (12.50)	34 (7.94)
	Total	144 (100.00)	156 (100.00)	128 (100.00)	428 (100.00)
Conventional Farmers					
1.	Illiterates	1 (1.35	2 (1.68)	2 (4.44)	5 (2.10)
2.	Primary	21 (28.38	32 (26.89)	6 (13.33)	59 (24.79)
3.	Secondary	32 (43.24	66 (55.46)	23 (51.11)	121 (50.84)
4.	Higher	13 (17.57	15 (12.61)	12 (26.67)	40 (16.81)
5.	Technical	7 (9.46	4 (3.36)	2 (4.44)	13 (5.46)

Note: Figures in parentheses indicate percentages to total.

Possession of farm assets may exert a considerable influence of farm activities. The farmer with own agricultural assets like farm machinery and livestock may perform his agricultural activities on time and similarly a farmer with sound financial assets in the form of gold and silver jewelry, deposits in financial institutions has more access to credit institutions and to other input markets. So in this regard, the asset structures of the selected farmers are computed on both per farm and per acre basis and the details are furnished in Table 4.8.

The analysis based on the per farm and per acre for paddy growing farmers (see Table 4.8.1) clearly indicated that there is no much difference between the organic farming category and conventional farming category with regard to the value of assets. The asset value per farm for paddy farmers in the organic farming category worked to ₹ 20.13 lakhs, where as for conventional farmers it is ₹ 20.42 lakhs. The same picture can be observed at the analysis of asset values per acre and the figure worked out to ₹ 2.53 lakhs for organic farming category and ₹ 2.48 lakhs for conventional category. It can also be observed that the farm size exhibits a positive relationship with asset value per farm and an inverse relationship with asset value per acre.

The analysis based on per farm and per acre for redgram growing farmers (Table 4.8.2) clearly indicates that there is not much difference between the organic farming category and conventional farming category with regard to the asset value. The asset value per farm for redgram farmers in the organic farming category is worked to be ₹ 49.40 lakhs where as for conventional farmers it is ₹ 51.41 lakhs. The same picture can be observed at the analysis of asset values per acre and the figures worked out to be ₹ 6.05 lakhs for organic farming category and ₹ 6.37 lakhs for conventional category. It can also be observed that the farm size exhibits a positive relationship with asset value per farm and negative relationship with asset value per acre.

The analysis based on the per farm and per acre for groundnut growing farmers (Table 4.8.3) clearly indicates that there is no much difference between the organic farming category and conventional farming category with regard to the asset value. The asset value per farm for groundnut farmers in the organic farming category is worked to be ₹ 58.38 lakhs where as for conventional farmers it is ‘ 52.87 lakhs. The same picture can be observed at the analysis of asset values per acre and the figures worked out to be ‘ 6.51 lakhs for organic farming category and ₹ 6.29 lakhs for conventional category. It can also be observed from the table that the farm size exhibits a positive relationship with asset value per farm and negative relationship with asset value per acre.

Table 4.8.1.1: Structure of Farm Assets (Per Farm) - Paddy Farmers (Organic and Conventional) (Value in '₹')

Organic Farmers					
S. No	*Asset Name*	*Small*	*Medium*	*Large*	*All Farms*
1.	Land	1164353	1342456	2319654	1466077
2.	Farm Buildings	38492	44652	164987	65658
Agri. Implements					
3.	Major	8541	14231	18145	12901
4.	Minor	4112	7924	11648	7246
Agri. Machinery					
5.	Tractor &Tractor Drawn Implements	98475	161358	235687	152671
6.	Live Stock	45268	75241	125975	74059
7.	Consumer Durables	59625	89463	112278	82933
8.	Financial Assets	112547	157863	215493	152389
9.	Total	1531413	1893188	3203867	2013935
Conventional Farmers					
S. No	*Asset Name*	*Small*	*Medium*	*Large*	*All Farms*
1.	Land	1321796	1451267	2092194	1561005
2.	Farm Buildings	31223	48957	123218	60606
Agri. Implements					
3.	Major	6751	11326	21315	12039
4.	Minor	3956	5003	8927	5576
Agri. Machinery					
5.	Tractor &Tractor Drawn Implements	82776	103699	154416	108218
6.	Live Stock	39216	44615	95019	55110
7.	Consumer Durables	62451	88761	106115	82839
8.	Financial Assets	115785	169214	203819	157028
9.	Total	1663954	1922842	2805023	2042421

Table 4.8.1.2: Structure of Farm Assets (Per Acre) - Paddy Farmers (Organic and Conventional) (Value in '₹')

Organic Farmers					
S. No	*Asset Name*	*Small*	*Medium*	*Large*	*All Farms*
1.	Land	356170	179866	129840	184722
2.	Farm Buildings	11775	5983	9235	8273
Agri. Implements					
3.	Major	2613	1907	1016	1626
4.	Minor	1258	1062	652	913
Agri. Machinery					
5.	Tractor &Tractor Drawn Implements	30123	21619	13192	19236
6.	Live Stock	13847	10081	7051	9331
7.	Consumer Durables	18239	11987	6285	10449
8.	Financial Assets	34428	21151	12062	19201
9.	Total	468452	253655	179332	253751
Conventional Farmers					
1.	Land	384989	188749	127139	189880
2.	Farm Buildings	9094	6367	7488	7372
Agri. Implements					
3.	Major	1966	1473	1295	1464
4.	Minor	1152	651	542	678
Agri. Machinery					
5.	Tractor &Tractor Drawn Implements	24110	13487	9384	13164
6.	Live Stock	11422	5803	5774	6704
7.	Consumer Durables	18190	11544	6448	10076
8.	Financial Assets	33724	22008	12386	19101
9.	Total	484647	250081	170456	248439

Table 4.8.2.1: Structure of Farm Assets (Per Farm) - Redgram Farmers (Organic and Conventional) (Value in '₹')

Organic Farmers					
S. No	*Asset Name*	*Small*	*Medium*	*Large*	*All Farms*
1.	Land	320238	717570	1464983	2388833
2.	Farm Buildings	24374	46732	89059	161605
Agri. Implements					
3.	Major	9268	22754	40373	70331
4.	Minor	6109	16260	23571	46229
Agri. Machinery					
5.	Tractor &Tractor Drawn Implements	44737	121633	221852	366388
6.	Live Stock	90557	186921	340834	620345
7.	Consumer Durables	51265	106999	301441	407558
8.	Financial Assets	102340	288089	566368	879034
9.	Total	648887	1506959	3048481	4940323
Conventional Farmers					
S. No	*Asset Name*	*Small*	*Medium*	*Large*	*All Farms*
1.	Land	379798	776768	1454259	2479856
2.	Farm Buildings	31249	49391	74188	164128
Agri. Implements					
3.	Major	7608	21040	35023	58698
4.	Minor	5335	9000	19553	32320
Agri. Machinery					
5.	Tractor &Tractor Drawn Implements	70095	130941	232402	425096
6.	Live Stock	113327	216498	396731	703665
7.	Consumer Durables	76729	144905	269684	475203
8.	Financial Assets	133715	246063	450388	810718
9.	Total	817856	1594605	2932227	5149686

Table 4.8.2.2: Structure of Farm Assets (Per Acre) - Redgram Farmers (Organic and Conventional) (Value in '₹')

Organic Farmers					
S. No	*Asset Name*	*Small*	*Medium*	*Large*	*All Farms*
1.	Land	103215	98456	91114	292785
2.	Farm Buildings	7856	6412	5539	19807
Agri. Implements					
3.	Major	2987	3122	2511	8620
4.	Minor	1969	2231	1466	5666
Agri. Machinery					
5.	Tractor &Tractor Drawn Implements	14419	16689	13798	44906
6.	Live Stock	29187	25647	21198	76032
7.	Consumer Durables	16523	14681	18748	49952
8.	Financial Assets	32985	39528	35225	107738
9.	Total	209141	206766	189599	605506
Conventional Farmers					
1.	Land	107635	102965	96541	307141
2.	Farm Buildings	8856	6547	4925	20328
Agri. Implements					
3.	Major	2156	2789	2325	7270
4.	Minor	1512	1193	1298	4003
Agri. Machinery					
5.	Tractor &Tractor Drawn Implements	19865	17357	15428	52650
6.	Live Stock	32117	28698	26337	87152
7.	Consumer Durables	21745	19208	17903	58856
8.	Financial Assets	37895	32617	29899	100411
9.	Total	231781	211374	194656	637811

Table 4.8.3.1: Structure of Farm Assets (Per Farm) - Groundnut Farmers (Organic and Conventional) (Value in '₹')

Organic Farmers					
S. No	*Asset Name*	*Small*	*Medium*	*Large*	*All Farms*
1.	Land	412683	115872	2097766	3103210
2.	Farm Buildings	34804	8756	146446	238764
Agri. Implements					
3.	Major	12043	2117	39224	69234
4.	Minor	4876	1219	17064	31815
Agri. Machinery					
5.	Tractor &Tractor Drawn Implements	60630	27415	654258	706214
6.	Live Stock	74746	19187	319476	518463
7.	Consumer Durables	64634	17456	219956	430515
8.	Financial Assets	131895	25615	342208	740378
9.	Total	796311	217637	3836398	5838593
Conventional Farmers					
S. No	*Asset Name*	*Small*	*Medium*	*Large*	*All Farms*
1.	Land	329256	804817	1897682	2761892
2.	Farm Buildings	28890	58155	109674	200737
Agri. Implements					
3.	Major	12091	24814	40217	81755
4.	Minor	4842	16272	31230	47146
Agri. Machinery					
5.	Tractor &Tractor Drawn Implements	56368	125582	313862	453954
6.	Live Stock	73654	160398	312747	541402
7.	Consumer Durables	66091	146726	344972	520407
8.	Financial Assets	81706	220134	413915	680020
9.	Total	652898	1556898	3464300	5287314

Table 4.8.3.2: Structure of Farm Assets (Per Acre) - Groundnut Farmers (Organic and Conventional) (Value in '₹')

Organic Farmers					
S. No	*Asset Name*	*Small*	*Medium*	*Large*	*All Farms*
1.	Land	123982	115872	106216	346070
2.	Farm Buildings	10456	8756	7415	26627
Agri. Implements					
3.	Major	3618	2117	1986	7721
4.	Minor	1465	1219	864	3548
Agri. Machinery					
5.	Tractor &Tractor Drawn Implements	18215	27415	33127	78757
6.	Live Stock	22456	19187	16176	57819
7.	Consumer Durables	19418	17456	11137	48011
8.	Financial Assets	39625	25615	17327	82567
9.	Total	239235	217637	194248	651120
Conventional Farmers					
1.	Land	112326	105897	110652	328875
2.	Farm Buildings	9856	7652	6395	23903
Agri. Implements					
3.	Major	4125	3265	2345	9735
4.	Minor	1652	2141	1821	5614
Agri. Machinery					
5.	Tractor &Tractor Drawn Implements	19230	16524	18301	54055
6.	Live Stock	25127	21105	18236	64468
7.	Consumer Durables	22547	19306	20115	61968
8.	Financial Assets	27874	28965	24135	80974
9.	Total	222737	204855	202000	629592

Summary

An analysis on demographic profile/characteristics has revealed that there is not much of difference in both the categories of farms viz., organic and conventional, like age, gender, family size etc., and economic characteristics like value of assets', size of land holding etc. Both the categories of farms can be differentiated with regard to the various levels of literacy, as the percentage of farmers with secondary and higher levels of education is more in organic farming category compared to their counterparts. As a result, it can be hypothesized that the farmers of organic farming category are more rational, have more accessibility to the information on organic farming practices, which consequently leads to efficient input-use.

Chapter 5

Costs of Cultivation and Returns From Organic *vis-a-vis* Conventional Farming

The earlier Chapter has dealt with the socio-economic profile of the sample households. In this Chapter an attempt has been made to analyse the cost of cultivation and returns from organic farming vis-à-vis conventional farming of the selected three crops. The standard concepts of costs and returns from farming of the Farm Management Studies (FMS)[1], sponsored by Directorate of Economics and Statistics, Ministry of Agriculture, Government of India, New Delhi, have been adopted in this Study and the results are analysed in Section - I. The perceptions of farmers on various issues relating to the organic farming have been presented in Section - II

SECTION - I

5.1 Cost of Cultivation

It is evident from the earlier studies that the cost of pesticides which constituted a major share in the total cost may be negligible for organic farming compared to the conventional farming, since organic pesticides are homemade and prepared with the locally available herbs. As a result, the organic farmers can get higher returns compared to their counterparts. In addition, chemical fertilisers are not supposed to be used in the case of organic farming. Though some other studies treated farm yard manure

(FYM) as a component of chemical fertilisers, this Study considered FYM as organic fertiliser. Except this minor difference, costs of remaining components that are necessary for calculating various cost concepts as per the Farm Management Studies (FMS) are used in totto. For studying all these aspects, an attempt has been made in this Chapter to compare the cost structure and returns from the selected crops for both the organic and conventional farming. Further, the economics of farm business for both organic and conventional farming has been analysed.

For studying the intensity of resource-use pattern, the total cost i.e., Cost – C has been adopted. Cost – C_2 is considered as the total cost and it includes the expenditure incurred on all the paid-out costs like seed, hired human labour, bullock labour(owned and hired), machine labour (owned and hired), farm yard manure (owned and purchased), chemical fertilizers, pesticides, irrigation charges, rent paid on leased-in land, etc., and imputed costs like depreciation on farm capital assets, interest on working capital, interest on farm fixed capital, rental value of owned land, imputed value of family labour etc. Though land revenue and cess have to be included in the total cost as per the standards, as the Govt. of Andhra Pradesh has stopped collection of these two they are not included in Cost - B. Different types of costs viz., Cost – A_1, Cost – A_2, Cost – B_1, Cost – B_2, Cost – C_1 and Cost – C_2 as used in the FMS are also computed and analysed. The details pertaining to these costs on the basis of per farm and per acre for organic and conventional farm holdings are presented in Tables - 5.1, 5.2 and 5.3.

It is evident from Table – 5.1.1 that the cost of paddy per acre on the basis of different cost concepts is found to be relatively higher on conventional farms compared to organic farms. The same phenomenon is discernible among different size groups of farms also. For instance, on the basis of Cost – A_1, the per acre cost on conventional farms is higher by 0.50 per cent, on the basis of Cost – C_2, this difference has gone up to 7.41 per cent. On organic farms, the proportions of Cost – A_1, Cost – A_2, Cost – B_1, Cost – B_2 and Cost – C_1 to total cost, i.e., Cost – C_2 worked out to about 63 per cent,63 per cent, 67 per cent, 98 per cent and 69 per cent respectively. Similarly, on conventional holdings these proportions worked out to about 59 per cent, 64 per cent, 62 per cent, 98 per cent and 64 per cent respectively. A similar pattern with variations in the proportions could also be observed among different size groups of farms. Further, it is to be noted that in case of organic holdings, the proportions of different costs to Cost – C_2 on one hand and farms size on the other hand are directly related, whereas in case of conventional holdings, the proportions of different costs to Cost – C_2 on one hand and farm size on the other hand are inversely related.

Table 5.1.1: Different Types of Costs of Cultivation Per Farm and Per Acre for Paddy (Value in '₹')

	Organic Farms		*Conventional Farms*	
Farm Size	*Per Farm*	*Per Acre*	*Per Farm*	*Per Acre*
Cost A_1				
Small	39786	12690	54094	15756
Medium	88027	12160	104461	13586
Large	236077	13731	215492	13095
All Farms	99057	13653	112576	13694
Cost A_2				
Small	39786	12690	56459	16445
Medium	88027	12160	119107	15491
Large	236077	13731	227822	13844
All Farms	99057	13653	121853	14822
Cost B_1				
Small	43917	13991	58570	17059
Medium	92746	12786	107865	14029
Large	249334	14473	228313	13874
All Farms	105211	14428	118752	14445
Cost B_2				
Small	63596	20190	83926	24445
Medium	144595	19660	180618	23491
Large	370068	21231	359470	21844
All Farms	158582	21153	187621	22822
Cost C_1				
Small	45431	14468	60716	17684
Medium	96057	13225	111700	14527
Large	253824	14724	233097	14165
All Farms	108091	14791	122165	14860
Cost C_2				
Small	65110	20667	86072	25070
Medium	147906	20099	184452	23989
Large	374558	21482	364253	22135
All Farms	161462	21515	191034	23237

Table 5.1.2: Different Types of Costs of Cultivation Per Farm and Per Acre for Redgram (Value in '₹')

	Organic Farms		*Conventional Farms*	
Farm Size	*Per Farm*	*Per Acre*	*Per Farm*	*Per Acre*
Cost A_1				
Small	16011	5173	21233	6017
Medium	44443	6091	49189	6520
Large	98420	6121	92845	6164
All Farms	48752	5975	50965	6312
Cost A_2				
Small	16011	5173	22111	6266
Medium	44443	6091	51805	6867
Large	98420	6121	100746	6688
All Farms	48752	5975	54258	6720
Cost B_1				
Small	17138	5538	22613	6409
Medium	45235	6199	49955	6622
Large	100061	6223	94805	6294
All Farms	49909	6117	52167	6461
Cost B_2				
Small	19879	6423	26522	7516
Medium	53565	7341	61235	8117
Large	118518	7371	119576	7938
All Farms	58951	7225	64350	7970
Cost C_1				
Small	18061	5836	23819	6750
Medium	47783	6548	52598	6972
Large	105903	6587	100093	6645
All Farms	52762	6467	54989	6811
Cost C_2				
Small	20802	6722	27727	7858
Medium	56113	7690	63877	8467
Large	124360	7735	124864	8289
All Farms	61804	7575	67172	8320

It is apparent from Table 5.1.2 that the cost of redgram per acre on the basis of different cost concepts is found to be relatively higher on conventional farms compared to organic farms. The same phenomenon is discernible among different size groups of farms also. For instance, on the basis of Cost – A_1, the per acre cost on conventional holdings is higher by five per cent compared to organic holdings. At the other extreme on the basis of Cost – C_2, this difference goes up to nine per cent. On organic holdings the proportion of cost – A_1, Cost – A_2, Cost – B_1, Cost – B_2 and Cost – C_1 to total cost, i.e., Cost – C_2 worked out to about 79 per cent, 79 per cent, 81 per cent, 95 per cent and 85 per cent respectively. Similarly, on conventional holdings these proportions worked out to about 76 per cent, 81 per cent, 78 per cent, 96 per cent and 82 per cent respectively. A similar pattern with variations in the proportions could also be observed among different size groups of farms. Further, it is to be noted that in case of organic holdings, the proportions of different costs to Cost – C_2 on one hand and farm-size on the other hand are directly related, whereas in case of conventional holdings, the proportions of different costs to Cost – C_2 on one hand and farms size on the other hand are inversely related.

It can be observed from Table 5.1.3 that the cost of groundnut per acre on the basis of different cost concepts is found to be relatively higher on conventional farms compared to organic farms. Except the small farm holdings, the same phenomenon is discernible among different size groups of farms also and the cost of cultivation for small farm holdings on organic farming is slightly higher. For instance, on the basis of Cost – A_1, the per acre cost on conventional holdings is higher by 17 per cent than that on organic holdings, while on the basis of Cost – C_2, this difference goes up to 18 per cent. On organic holdings the proportion of cost – A_1, Cost – A_2, Cost – B_1, Cost – B_2 and Cost – C_1 to total cost, i.e., Cost – C_2 worked out to about 91 per cent, 91 per cent, 94 per cent, 96 per cent and 98 per cent respectively. Similarly, on conventional holdings these proportions worked out to about 90 per cent, 91 per cent, 92 per cent, 96 per cent and 96 per cent respectively. A similar pattern with variations in the proportions could also be observed among different size groups of farms also. Further, it is to be noted that in case of organic holdings, the proportions of different costs to Cost – C_2 on one hand and farms size on the other hand are inversely related, whereas in case of conventional holdings, the proportions of different costs to Cost – C_2 on one hand and farms size on the other hand are directly related.

Table 5.1.3: Different Types of Costs of Cultivation Per Farm and Per Acre for Groundnut (Value in '₹')

	Organic Farms		*Conventional Farms*	
Farm Size	*Per Farm*	*Per Acre*	*Per Farm*	*Per Acre*
Cost A_1				
Small	61945	18609	52920	18054
Medium	125901	16857	136491	17959
Large	278275	14091	337333	19670
All Farms	140088	15622	157950	18808
Cost A_2				
Small	61945	18609	53393	18215
Medium	125901	16857	138461	18219
Large	278275	14091	342314	19960
All Farms	140088	15622	160164	19072
Cost B_1				
Small	65819	19773	55812	19040
Medium	133816	17917	138608	18238
Large	283837	14373	344081	20063
All Farms	146024	16284	161427	19222
Cost B_2				
Small	65274	19609	56324	19215
Medium	133369	17857	146061	19219
Large	298025	15091	359464	20960
All Farms	149055	16622	168562	20072
Cost C_1				
Small	68279	20512	58273	19880
Medium	138815	18586	144882	19063
Large	294814	14929	359510	20963
All Farms	151569	16902	168678	20086
Cost C_2				
Small	67734	20348	58785	20055
Medium	138368	18526	152336	20044
Large	309002	15647	374893	21860

5.2 Resource Use Pattern

To ascertain the relative importance of different inputs in the cost structure, an item-wise breakup of the total cost is computed. The details for organic and conventional holdings on the basis of per farm and per acre for different size groups of farms are presented in Table 5.2.

It can be observed from the Table that the total cost per acre on organic farm holdings of the three selected crops viz., paddy, redgram and groundnut worked out to ₹ 21,549/-, ₹ 7,717/- and ₹ 17,903/- respectively, whereas on conventional holding these values are worked out to be ₹ 23,989/-, ₹ 8,468/- and ₹ 21,349/- which clearly showed that the cost of cultivation for conventional holdings is higher by 11 per cent, 10 per cent and 19 per cent compared to organic farming households for the three selected crops respectively.

Among different inputs, hired human labour, machine labour, farmyard manure, pesticides, seed and bullock labour appeared to be predominant in the cost structure for both Organic and Conventional farms, for all the three selected crops. In addition, fertiliser appeared to be predominant in conventional farms only.

In case of organic paddy farms, apart from the imputed costs, the proportion of expenditure incurred on human labour accounts for about 32 per cent of the total cost (Table 5.2.1). This is followed by the proportion of expenditure incurred on organic fertiliser (10 per cent), machine labour (8 per cent), pesticide (2 per cent), seed (2 per cent) etc. A similar pattern with minor variations in the proportions could be observed among different size groups of farms. It could be also observed that the proportion of expenditure on human labour to total cost has exhibited a direct relationship with farm size. As far as the cost structure of the organic redgram farms is concerned (Table 5.2.3), again the expenditure on human labour appeared to be predominant (30 per cent) and this is followed by organic fertiliser (14 per cent), pesticides (8 per cent), bullock labour (7 per cent), machine labour (3 per cent) and seed (2 per cent).

Table 5.2.1.1: Cost of Cultivation of Paddy Per Farm - Organic Farms

(Value in '₹')

		Small	*Medium*	*Large*	*All Farms*
1	Human Labour	19143	44940	143450	54526
2	Bullock Labour	1222	244	2237	988
3	Machine Labour	5007	14203	29414	13772
4	Seed	1444	3588	8073	3669
5	Organic Fertilisers	7144	16690	36762	17070
6	Pesticides	1481	4049	7276	3731
7	Others	1512	4121	6162	3559
8	Interest on working capital	2309	1925	4380	2540
9	Depreciation	1982	2322	4455	2610
10	Rent Paid on Leased-in land	0	0	0	0
11	Interest on Fixed Capital	4131	4719	13257	6154
12	Rental Value of Owned Land	23809.09	56568.18	133991.38	59525
13	Imputed Value of Family Labour	1514	3311	4490	2880
	Total	70697	156680	393949	171025

Table 5.2.1.2: Cost of Cultivation of Paddy Per Acre - Organic Farms

(Value in '₹')

		Small	*Medium*	*Large*	*All Farms*
1	Human Labor	6030 (27.08)	5958 (28.68)	8029 (36.41)	6870 (31.88)
2	Bullock Labour	385 (1.73)	32 (0.16)	125 (0.57)	124 (0.58)
3	Machine Labour	1577 (7.08)	1883 (9.06)	1646 (7.47)	1735 (8.05)
4	Seed	455 (2.04)	476 (2.29)	452 (2.05)	462 (2.15)
5	Organic Fertilisers	2250 (10.11)	2213 (10.65)	2058 (9.33)	2151 (9.98)
6	Pesticides	466 (2.09)	537 (2.58)	407 (1.85)	470 (2.18)
7	Others	476 (2.14)	546 (2.63)	345 (1.56)	448 (2.08)
8	Interest on working capital	728 (3.27)	255 (1.23)	245 (1.11)	320 (1.49)
9	Depreciation	624 (2.80)	308 (1.48)	249 (1.13)	329 (1.53)
10	Rent Paid on Leased-in land	0 (0.00)	0 (0.00)	0 (0.00)	0 (0.00)
11	Interest on Fixed Capital	1301 (5.84)	626 (3.01)	742 (3.37)	775 (3.60)
12	Rental Value of Owned Land	7500 (33.68)	7500 (36.10)	7500 (34.01)	7500 (34.80)
13	Imputed Value of Family Labour	477 (2.14)	439 (2.11)	251 (1.14)	363 (1.68)
	Total	22270 (100.00)	20773 (100.00)	22051 (100.00)	21549 (100.00)

Table 5.2.2.1: Cost of Cultivation of Paddy Per Farm - Conventional Farms

(Value in '₹')

		Small	*Medium*	*Large*	*All Farms*
1	Human Labour	27229	50447	108895	56004
2	Bullock Labour	1640	539	2111	1361
3	Machine Labour	6591	14411	31430	15616
4	Seed	2014	3915	8524	4326
5	Chemical Fertilisers	6225	13645	29490	14712
6	Pesticides	2871	5251	6467	4627
7	Others	2142	5139	5108	3963
8	Interest on working capital	3045	5834	12002	6288
9	Depreciation	2336	5280	11466	5678
10	Rent Paid on Leased-in land	2365	14646	12330	9278
11	Interest on Fixed Capital	4476	3404	12822	6177
12	Rental Value of Owned Land	27466	61511	131648	65768
13	Imputed Value of Family Labour	2146	3834	4784	3413
	Total	990548	187856	377075	197211

With regard to organic groundnut farms (Table 5.2.5), again, the expenditure on human labour constitutes about 38 per cent of the total cost and it is followed by seed (12 per cent), bullock labour (8 per cent), organic fertiliser (7 per cent), pesticides (6 per cent) and machine labour (2 per cent).

On the other hand, in case of conventional farms of the three selected crop also, the proportion of expenditure to total cost incurred on human labour is the highest - 28 per cent, 29 per cent and 34 per cent for paddy, redgram and groundnut respectively(Tables – 5.2.2, 5.2.4 and 5.2.6). With regard to the other components of total cost for conventional paddy farms, the expenditure on human labour is followed by machine labour (8 per cent), fertilisers (6 per cent), pesticides (2 per cent), seed (2 per cent) and

Table 5.2.2.2: Cost of Cultivation of Paddy Per Acre - Conventional Farms

(Value in '₹')

		Small	*Medium*	*Large*	*All Farms*
1	Human Labour	7931 (30.07)	6561 (26.85)	6617 (28.88)	6812 (28.40)
2	Bullock Labour	478 (1.81)	70 (0.29)	128 (0.56)	166 (0.69)
3	Machine Labour	1920 (7.28)	1874 (7.67)	1910 (8.34)	1900 (7.92)
4	Seed	587 (2.22)	509 (2.08)	518 (2.26)	526 (2.19)
6	Chemical Fertilisers	1813 (6.88)	1774 (7.26)	1792 (7.82)	1790 (7.46)
7	Pesticides	836 (3.17)	683 (2.80)	393 (1.72)	563 (2.35)
8	Others	624 (2.37)	668 (2.74)	310 (1.35)	482 (2.01)
9	Interest on working capital	887 (3.36)	759 (3.11)	729 (3.18)	765 (3.19)
10	Depreciation	680 (2.58)	687 (2.81)	697 (3.04)	691 (2.88)
11	Rent Paid on Leased-in land	689 (2.61)	1905 (7.80)	749 (3.27)	1129 (4.70)
12	Interest on Fixed Capital	1304 (4.94)	443 (1.81)	779 (3.40)	751 (3.13)
13	Rental Value of Owned Land	8000 (30.33)	8000 (32.74)	8000 (34.91)	8000 (33.35)
14	Imputed Value of Family Labour	625 (2.37)	499 (2.04)	291 (1.27)	415 (1.73)
	Total	26373 (100.00)	24432 (100.00)	22914 (100.00)	23989 (100.00)

farm yard manure (2 per cent) etc. With regard to the conventional redgram farms, the expenditure on human labour is followed by fertiliser (11 per cent), pesticides (7 per cent), bullock labour (6 per cent), farmyard manure (4 per cent), machine labour (3 per cent) and seed (2 per cent). With regard to the conventional groundnut farms, the expenditure on human labour is followed by pesticides (12 per cent), seed (11 per cent), bullock labour (7 per cent), fertiliser (5 per cent) machine labour (4 per cent) and farm yard manure (1 per cent).

The above analysis has revealed that the proportion of expenditure on organic fertilisers is higher for organic paddy farms when compared with the expenditure on fertilisers on conventional paddy farms. However, the total cost per acre on organic farms is lower than that on conventional farms due to the lower Table 5.2.3.1Cost of Cultivation of Redgram Per Farm - Organic Farms(Value in '₹') expenditure on other inputs. A similar picture with slight variations in proportions can be observed with regard to the redgram and groundnut also.

Table 5.2.3.1: Cost of Cultivation of Redgram Per Farm - Organic Farms

(Value in ₹

		Small	*Medium*	*Large*	*All Farms*
1	Human Labour	6519	18009	37752	19171
2	Bullock Labour	1384	3823	8763	4279
3	Machine Labour	615	1699	3895	1902
4	Seed	369	1019	2335	1141
5	Organic Fertilisers	2767	7645	17526	8558
6	Pesticides	1537	4247	9737	4755
7	Others	1384	3823	8763	4279
8	Interest on working capital	911	2517	5548	2755
9	Depreciation	525	1661	4101	1912
10	Rent Paid on Leased-in land	0	0	0	0
11	Interest on Fixed Capital	1127	791	1642	1157
12	Rental Value of Owned Land	3868	9121	20098	10199
13	Imputed Value of Family Labour	923	2548	5842	2853
	Total	21929	56904	126002	62961

Table 5.2.3.2: Cost of Cultivation of Redgram Per Acre – Organic Farms

(Value in ₹')

		Small	*Medium*	*Large*	*All Farms*
1.	Human Labour	2106 (29.73)	2468 (31.65)	2348 (29.96)	2350 (30.45)
2.	Bullock Labour	447 (6.31)	524 (6.72)	545 (6.95)	524 (6.80)
3.	Machine Labour	199 (2.80)	233 (2.99)	242 (3.09)	233 (3.02
4.	Seed	119 (1.68)	140 (1.79)	145 (1.85)	140 (1.81)
5.	Organic Fertilisers	894 (12.62)	1048 (13.44)	1090 (13.91)	1049 (13.59)
6.	Pesticides	497 (7.01)	582 (7.46)	606 (7.73)	583 (7.55)
7.	Others	447 (6.31)	524 (6.72)	545 (6.95)	524 (6.80)
8.	Interest on working capital	294 (4.15)	345 (4.42)	345 (4.40)	338 (4.38)
9.	Depreciation	170 (2.39)	228 (2.92)	255 (3.25)	234 (3.04)
10.	Rent Paid on Leased-in land	0 (0.00)	0 (0.00)	0 (0.00)	0 (0.00)
11.	Interest on Fixed Capital	364 (5.14)	108 (1.39)	102 (1.30)	142 (1.84)
12.	Rental Value of Owned Land	1250 (17.64)	1250 (16.03)	1250 (15.95)	1250 (16.20)
13.	Imputed Value of Family Labour	298 (4.21)	349 (4.48)	363 (4.64)	350 (4.53)
	Total	7086 (100)	7798 (100)	7837 (100)	7717 (100)

Table 5.2.4.1: Cost of Cultivation of Redgram Per Farm – Conventional Farms

(Value in '₹')

		Small	*Medium*	*Large*	*All Farms*
1	Human Labour	8329	18672	36108	19612
2	Bullock Labour	1749	3963	7932	4217
3	Machine Labour	781	1761	3524	1875
4	Seed	471	1057	2095	1121
6	Chemical Fertilisers	4189	10770	17544	10417
7	Pesticides	1954	4404	8814	4688
8	Others	1749	3963	7932	4217
9	Interest on working capital	1201	2787	5247	2884
10	Depreciation	810	1812	3649	1936
11	Rent Paid on Leased-in land	879	2616	7902	3292
12	Interest on Fixed Capital	1381	767	1960	1201
13	Rental Value of Owned Land	4411	9430	18830	10093
14	Imputed Value of Family Labour	1204	2642	5288	2822
	Total	29108	64643	126825	68373

Table 5.2.4.2: Cost of Cultivation of Redgram Per Acre – Conventional Farms

(Value in '₹')

		Small	*Medium*	*Large*	*All Farms*
1	Human Labour	2360 (28.61)	2475 (28.88)	2397 (28.47)	2429 (28.68)
2	Bullock Labour	496 (6.01)	525 (6.13)	527 (6.25)	522 (6.17)
3	Machine Labour	221 (2.68)	233 (2.72)	234 (2.78)	232 (2.74)
4	Seed	133 (1.62)	140 (1.64)	139 (1.65)	139 (1.64)
5	Chemical Fertilisers	1187 (14.39)	1428 (16.66)	1165 (13.83)	1291 (15.23)
6	Pesticides	554 (6.71)	584 (6.81)	585 (6.95)	581 (6.86)
7	Others	496 (6.01)	525 (6.13)	527 (6.25)	522 (6.17)
8	Interest on working capital	340 (4.13)	369 (4.31)	348 (4.14)	357 (4.22)
9	Depreciation	230 (2.78)	240 (2.80)	242 (2.88)	240 (2.83)
10	Rent Paid on Leased-in land	249 (3.02)	347 (4.05)	525 (6.23)	408 (4.82)
11	Interest on Fixed Capital	391 (4.74)	102 (1.19)	130 (1.55)	149 (1.76)
12	Rental Value of Owned Land	1250 (15.15)	1250 (14.59)	1250 (14.85)	1250 (14.76)
13	Imputed Value of Family Labour	342 (4.14)	350 (4.09)	351 (4.17)	350 (4.13)
	Total	8249 (100.00)	8569 (100.00)	8419 (100.00)	8468 (100.00)

A regression equation of the form of $\ln L = \ln a + b \ln X$ (where L = value of hired human labour input per acre and X = farm size in acres) is fitted to examine the relationship between farm size and labour-use. The fitted regression equations are:

Paddy:

$\ln L = 8.73 + 0.46 \ln X$ ——— Organic farms

$\ln L = 8.15 + 0.38 \ln X$ ——— Conventional farms

Redgram:

$\ln L = 0.91 + 0.19 \ln X$ ——— Organic farms

$\ln L = 1.79 + 0.16 \ln X$ ——— Conventional farms

Groundnut:

$\ln L = 3.68 + 0.32 \ln X$ ——— Organic farms

$\ln L = 2.11 + 0.24 \ln X$ ——— Conventional farms

The results of the regression equations have showed that the relationship between farm size and expenditure on hired human labour is positive, which is in conformity with our earlier observation. In case of organic farms this coefficient is found to be significant at 5 per cent level, while it is significant at 5 to 10 per cent probability levels on Conventional farms.

In order to examine the relationship between farm size (X) and expenditure on fertilizer per acre (F), regression equations are estimated for both organic and conventional farms. The estimated regression equations are:

$\ln L = 3.68 + 0.32 \ln X$ ——— Organic farms

Paddy

$\ln F = 7.13 - 0.43 \ln X$ ——— Organic farms

$\ln F = 6.71 - 0.28 \ln X$ ——— Conventional farms

Table 5.2.5.1: Cost of Cultivation of Groundnut Per Farm - Organic Farms

(Value in '₹')

		Small	*Medium*	*Large*	*All Farms*
1	Human Labour	26650	54151	118915	60069
2	Bullock Labour	5945	12080	26527	13400
3	Machine Labour	1435	2916	6403	3235
4	Seed	8815	17911	39333	19869
5	Organic Fertilisers	5125	10414	22868	11552
6	Pesticides	4613	9372	20581	10397
7	Others	3916	7956	17471	8826
8	Interest on working capital	3531	7175	15756	7959
9	Depreciation	1915	3926	10421	4781
10	Rent Paid on Leased-in land	0	0	0	0
11	Interest on Fixed Capital	3874	7915	5562	5936
12	Rental Value of Owned Land	3329	7468	19750	8967
13	Imputed Value of Family Labour	2460	4999	10977	5545
	Total	71607	146282	314565	160534

Table 5.2.5.2: Cost of Cultivation of Groundnut Per Acre - Organic Farms

(Value in ₹')

		Small	*Medium*	*Large*	*All Farms*
1	Human Labour	8006 (37.22)	7251 (37.02)	6021 (37.80)	6699 (37.42)
2	Bullock Labour	1786 (8.30)	1617 (8.26)	1343 (8.43)	1494 (8.35)
3	Machine Labour	431 (2.00)	390 (1.99)	324 (2.04)	361 (2.01)
4	Seed	2648 (12.31)	2398 (12.24)	1992 (12.50)	2216 (12.38)
5	Organic Fertilisers	1540 (7.16)	1394 (7.12)	1158 (7.27)	1288 (7.20)
6	Pesticides	1386 (6.44)	1255 (6.41)	1042 (6.54)	1159 (6.48)
7	Others	1176 (5.47)	1065 (5.44)	885 (5.55)	984 (5.50)
8	Interest on working capital	1061 (4.93)	961 (4.90)	798 (5.01)	888 (4.96)
9	Depreciation	575 (2.67)	526 (2.68)	528 (3.31)	533 (2.98)
10	Rent Paid on Leased-in land	0 (0.00)	0 (0.00)	0 (0.00)	0 (0.00)
11	Interest on Fixed Capital	1164 (5.41)	1060 (5.41)	282 (1.77)	662 (3.70)
12	Rental Value of Owned Land	1000 (4.65)	1000 (5.11)	1000 (6.28)	1000 (5.59)
13	Imputed Value of Family Labour	739 (3.44)	669 (3.42)	556 (3.49)	618 (3.45)
	Total	21513 (100.00)	19587 (100.00)	15927 (100.00)	17903 (100.00)

Table 5.2.6.1: Cost of Cultivation of Groundnut Per Farm - Conventional Farms

(Value in '₹')

		Small	*Medium*	*Large*	*All Farms*
1	Human Labour	20846	53147	130690	61421
2	Bullock Labour	3909	9965	24505	11516
3	Machine Labour	2389	6090	14975	7038
4	Seed	6804	17347	42656	20047
5	Chemical Fertilisers	4059	10349	25448	11960
6	Pesticides	7470	19044	46831	22009
7	Others	3257	8304	20420	9597
8	Interest on working capital	3046	7765	19095	8974
9	Depreciation	1141	4480	12713	5387
10	Rent Paid on Leased-in land	473	1971	4981	2214
11	Interest on Fixed Capital	2892	2117	6748	3477
12	Rental Value of Owned Land	2931	7600	17150	8398
13	Imputed Value of Family Labour	2461	6274	15429	7251
	Total	61677	154453	381641	179290

Table 5.2.6.2: Cost of Cultivation of Groundnut Per Acre - Conventional Farms

(Value in ₹)

		Small	*Medium*	*Large*	*All Farms*
1	Human Labour	7111 (33.8)	6993 (34.41)	7620 (34.24)	7314 (34.26)
2	Bullock Labour	1333 (6.34)	1311 (6.45)	1429 (6.42)	1371 (6.42)
3	Machine Labour	815 (3.87)	801 (3.94)	873 (3.92)	838 (3.93)
4	Seed	2321 (11.03)	2282 (11.23)	2487 (11.18)	2387 (11.18)
5	Chemical Fertilisers	1385 (6.58)	1362 (6.70)	1484 (6.67)	1424 (6.67)
6	Pesticides	2548 (12.11)	2506 (12.33)	2731 (12.27)	2621 (12.28)
7	Others	1111 (5.28)	1093 (5.38)	1191 (5.35)	1143 (5.35)
8	Interest on working capital	1039 (4.94)	1022 (5.03)	1113 (5.00)	1069 (5.01)
9	Depreciation	389 (1.85)	589 (2.90)	741 (3.33)	642 (3.00)
10	Rent Paid on Leased-in land	161 (0.77)	259 (1.28)	290 (1.31)	264 (1.23)
11	Interest on Fixed Capital	987 (4.69)	279 (1.37)	393 (1.77)	414 (1.94)
12	Rental Value of Owned Land	1000 (4.75)	1000 (4.92)	1000 (4.49)	1000 (4.68)
13	Imputed Value of Family Labour	840 (3.99)	826 (4.06)	900 (4.04)	863 (4.04)
	Total	21041 (100.00)	20323 (100.00)	22253 (100.00)	21349 (100.00)

Redgram:

$\ln F = 3.68 + 0.13 \ln X$ ———— Organic farms

$\ln F = 1.79 - 0.08 \ln X$ ———— Conventional farms

Groundnut:

$\ln F = 4.13 - 0.17 \ln X$ ———— Organic farms

$\ln F = 3.19 + 0.27 \ln X$ ———— Conventional farms

The results have indicated an inverse relationship exists between farm size and per acre expenditure on fertilizers on both organic and conventional farms with the exception of organic redgram farms and conventional groundnut farms. In both of these, a positive relationship is exhibited. However, while the coefficient associated with this variable (F) is found to be significant at one per cent probability level in case of organic farms, it is significant at 10 per cent probability level in case of conventional farms. These findings also collaborates the earlier observations of tabular analysis.

Similarly, a regression equation is fitted between farms size(X) and per acre expenditure on pesticides (P). The estimated regression equations are:

Paddy:

$\ln P = 9.17 - 0.98 \ln X$ ———— Organic farms

$\ln P = 9.87 - 0.89 \ln X$ ———— Conventional farms

Redgram:

$\ln P = 6.97 + 0.18 \ln X$ ———— Organic farms

$\ln P = 4.39 + 0.37 \ln X$ ———— Conventional farms

Groundnut:

$\ln P = - 0.69 + 0.08 \ln X$ ———— Organic farms

$\ln P = 1.57 + 0.11 \ln X$ ———— Conventional farms

The results have indicated a positive and significant relationship between farm size and per acre expenditure on pesticides in case of redgram and groundnut for both organic and conventional farms. On the other hand, in case of paddy, an inverse and significant relationship is found between farm size and per acre expenditure on pesticides.

Finally, to examine the relationship between farm size and total cost (Cost - C_2) per acre, the regression equation of the form $\ln C = \ln a + b \ln X$ is estimated and the estimated regression equations are:

Paddy:

$\ln C = 11.43 - 0.53 \ln X$ ———— Organic farms

$\ln C = 11.54 + 0.39 \ln X$ ———— Conventional farms

Redgram:

$\ln C = 14.97 + 0.39 \ln X$ ———— Organic farms

$\ln C = 17.81 + 0.26 \ln X$ ———— Conventional farms

Groundnut:

$\ln C = 11.89 + 0.27 \ln X$ ———— Organic farms

$\ln C = 10.17 + 0.18 \ln X$ ———— Conventional farms

Both the regression coefficients are found to be significant at probability levels ranging from one to 10 per cent, indicating a direct relationship between farm size and total cost with the exception of organic paddy, where-in an inverse relationship is exhibited between these two.

5.3 Different Types of Costs

A break-up of total cost into different types of costs viz., Prime cost, Operational cost, Overhead cost, Paid-out cost and Imputed cost is must for wise analysis of cost of cultivation in agricultural economic studies. Hence, an attempt is made in this direction and the proportion of different types of costs to total cost has been computed for all the three selected crops and presented in Table 5.3.

5.3.1 Prime Cost

As cost – C_2 also includes imputed values; it may not represent the true cost of cultivation of the farmer. As prime cost includes all paid out expenses (represented by Cost – A) and value of family labour excluding irrigation charges, it was considered relevant for the purpose of the Study. It can be observed from Table 5.3.1 that the proportion of prime cost per farm and per acre on organic paddy farms to total cost is around 83 per cent and on conventional farms it is around 81 per cent. With slight changes the same picture can be found in the case of small farmers. In the case of medium farmers the prime cost per acre is around 6 per cent higher for organic farm households than the conventional farm households. For large farmers, the difference is only around 2 per cent.

With regard to redgram, it can be seen from Table 5.3.2 that the proportion of prime cost per acre and per farm on organic farms to total cost is around 81 per cent and on conventional farms it is 83 per cent. A size-wise analysis shows that the prime cost per acre and per farm on organic farms for small, medium and large farms is around 81 per cent, 78 per cent and 81 per cent respectively, while the same on conventional farms are 82 per cent, 81 per cent and 81 per cent respectively. Further, prime cost per acre on conventional farms is consistently higher than that on organic farms.

With regard to the groundnut, it is apparent from Table 5.3.3, that the proportion of prime cost per farm and per acre on organic farms to total cost is around 78 per cent and on conventional farms it is around 79 per cent implying that the prime cost on organic groundnut farms is 0.9 per cent lower than the conventional groundnut farms. In the case of small farms, the prime cost is about five percent higher on conventional farms than on the organic farms. In the case of medium and large farms, the prime cost per farm and per acre is around 2 per cent and 0.56 per cent respectively higher on conventional farms than the organic farms.

5.3.2 Operational Cost

Operational cost includes expenditure on seeds, fertilizers, manures, pesticides, hired human labour, bullock labour and machine labour. With regard to paddy farms, it can be found (Table - 5.3.1) that the proportion of operational cost to total cost per acre is 65.41 per cent on organic farms and 61.51 per cent on the conventional farms. With regard to small farmers, the operational cost per acre for Organic farmers is 61 per cent and for Conventional farmers it is 59 per cent. With regard to medium farmers, it is 67 per cent on organic farms and 61 per cent on conventional farms. With regard to large farmers, it is 67 per cent and 66 per cent respectively for organic and conventional farms. It can be concluded that the proportion of operational cost to total cost is lower on conventional farms than the organic farms both at the aggregate level (all farms) and disaggregate level (by size-wise).

As far as the redgram farming is considered, (Table - 5.3.2) the proportion of operational cost to total cost is around 3 per cent higher on conventional farms than the organic farms. With regard to different farm size groups, the conventional farms recorded a higher proportion of operational cost to total cost by 0.39 per cent for small farms, 5.17 per cent for medium farms and 4.53 per cent for large farms than the organic farms.

As far as the groundnut farming is considered, (Table - 5.3.3) the proportion of operational cost to total cost is slightly more than 4 per cent higher on conventional farms compared to the organic farms. With regard to different size groups of farms, the conventional farms recorded higher operational cost per acre than the corresponding size groups of organic farming.

5.3.3 Overhead Costs

The expenses like depreciation, water taxes, rent paid for the leased-in land and rental value of owned land come under overhead costs. It can be observed that the proportion of overhead cost to total cost is around 3 per cent higher on conventional paddy farms than the organic paddy farms. With regard to different size groups of farms, the conventional farms recorded higher proportion of overhead cost to total cost by 2.81 per cent for small farms, 3.81 per cent for medium farms and 2.01 per cent for large farms than the order farm to total cost is higher on conventional farms by 1.60 per cent than the organic farms. With regard to different size groups of farms, the conventional farms registered higher proportion of overhead cost to total cost by 2.58 per cent for small farms, 3.51 per cent for medium farms and 2.50 per cent for large farms than the organic farms.

5.3.4 Paid-out Cost

One of the indicators of progressive farming in agriculture is higher proportion of paid out costs in total cost. Paid-out cost includes all the out of pocket expenses incurred by the farmer on seed, fertilizers, pesticides, hired human labour and other inputs. As far as the paddy farming is considered, it is evident from Table - 5.3.1 that the proportion of paid out cost to total cost per acre and per farm is higher for conventional farms than the organic farms and the difference is about 7 per cent. With regard to different size groups of farms, the proportion of paid out cost to total cost is much higher for medium and large conventional farms and the difference is about 10 per cent. With regard to small farms, the difference between conventional and organic farms in relation to the proportion of paid out cost to total cost is much lesser than other size groups of farms and it is around 2 per cent.

As far as the redgram farming is considered, Table - 5.3.2 shows that the proportion of paid out cost to total cost per acre and per farm on conventional farms is higher by about 5 per cent than organic farms. With regard to different size groups of farms, medium farms registered a higher proportion of paid out cost to total cost and the difference is 9 per cent, whereas the small and large farms registered a lower proportion and the difference is marginal.

As far as the groundnut farming is considered, it can be observed from Table - 5.3.3, that the proportion of paid out cost to total cost per acre and per farm on conventional farms is higher by about 4 per cent than the conventional farms. With regard to different size groups of farms, the difference between conventional and organic farms in relation to the proportion of paid out cost to total cost, is much higher for medium farms (11 per cent) and lower for small and large farms (2 per cent and 1 per cent respectively).

5.3.5 Imputed Cost

The imputed cost on the other hand, includes imputed values of inputs owned by the farmer such as depreciation, interest on fixed capital, rental value of owned land and the value of family labour. As far as the paddy farming is considered, Table - 5.3.1 reveals that the proportion of imputed cost to total cost per acre and per farm is higher on conventional farms than the organic farms and the difference is about 3 per cent. With regard to different size groups of farms, the difference is much higher for medium farms (8 per cent) and lesser for small and large farms (about one per cent).

Table 5.3.1: Different Types of Costs Per Farm and Per Acre for Paddy

(Value in '₹')

	Organic Farms		*Conventional Farms*	
Farm Size	*Per Farm*	*Per Acre*	*Per Farm*	*Per Farm*
Prime Cost				
Small	53468 (82.12)	16972 (82.12)	69228 (80.43)	20164 (80.43)
Medium	123383 (83.42)	16767 (83.42)	143153 (77.61)	18618 (77.61)
Large	307550 (82.11)	17639 (82.11)	308413 (84.67)	18742 (84.67)
All Farms	133238 (82.52)	17754 (82.52)	155043 (81.16)	18859 (81.16)
Operational Cost				
Small	39411 (60.53)	12510 (60.53)	50782 (59.00)	14791 (59.00)
Medium	98949 (66.90)	13446 (66.90)	111686 (60.55)	14525 (60.55)
Large	252340 (67.37)	14472 (67.37)	231301 (63.50)	14498 (65.50)
All Farms	105612 (65.41)	14073 (65.41)	117505 (61.51)	14293 (61.51)
Overhead Cost				
Small	18205 (27.96)	5778 (27.96)	26484 (30.77)	7714 (30.77)
Medium	38855 (26.27)	5280 (26.27)	55483 (30.08)	7216 (30.08)
Large	103078 (27.52)	5912 (27.52)	107564 (29.53)	6536 (29.53)
All Farms	44031 (27.27)	5867 (27.27)	57272 (29.98)	6966 (29.98)
Paid-out Cost				
Small	43324 (66.54)	13752 (66.54)	59192 (68.77)	17241 (68.77)
Medium	97041 (65.61)	13187 (65.61)	139852 (75.82)	18188 (75.82)
Large	245748 (65.61)	14094 (65.61)	275193 (75.55)	16723 (75.55)
All Farms	107760 (66.74)	14359 (66.74)	141040 (73.83)	17156 (73.83)
Imputed Costs				
Small	17690 (27.17)	5615 (27.17)	24384 (28.33)	7102 (28.33)
Medium	31445 (21.26)	4273 (21.26)	53989 (29.27)	7022 (29.27)
Large	75623 (20.19)	4337 (20.19)	73871 (20.28)	4489 (20.28)
All Farms	36119 (22.37)	4813 (22.37)	48255 (25.26)	5870 (25.26)

Table 5.3.2: Different Types of Costs Per Farm and Per Acre for Redgram

(Value in '₹')

	Organic Farms		*Conventional Farms*	
Farm Size	*Per Farm*	*Per Acre*	*Per Farm*	*Per Acre*
Prime Cost				
Small	16920 (81.34)	5468 (81.34)	22814 (82.28)	6466 (82.28)
Medium	43981 (78.38)	6027 (78.38)	51919 (81.28)	6882 (81.28)
Large	100632 (80.92)	6259 (80.92)	101739 (81.48)	6754 (81.48)
All Farms	50321 (81.42)	6168 (81.42)	55968 (83.32)	6932 (83.32)
Operational Cost				
Small	12229 (58.79)	3952 (58.79)	16409 (59.18)	4650 (59.18)
Medium	33735 (60.12)	4623 (60.12)	41705 (65.29)	5528 (65.29)
Large	79939 (64.28)	4972 (64.28)	85919 (68.81)	5704 (68.81)
All Farms	37651 (60.92)	4615 (60.92)	43151 (64.24)	5345 (64.24)
Overhead Cost				
Small	5446 (26.18)	1760 (26.18)	7938 (28.63)	2250 (28.63)
Medium	14657 (26.12)	2009 (26.12)	19681 (30.81)	2609 (30.81)
Large	33478 (26.92)	2082 (26.92)	35399 (28.35)	2350 (28.35)
All Farms	17423 (28.19)	2135 (28.19)	18325 (27.28)	2270 (27.28)
Paid out Cost				
Small	14170 (68.12)	4579 (68.12)	19209 (69.28)	5444 (69.28)
Medium	35407 (63.10)	4852 (63.10)	46106 (72.18)	6111 (72.18)
Large	94402 (75.91)	5872 (75.91)	95046 (76.12)	6310 (76.12)
All Farms	42107 (68.13)	5161 (68.13)	48975 (72.91)	6066 (72.91)
Imputed Costs				
Small	5433 (26.12)	1756 (26.12)	7539 (27.19)	2137 (27.19)
Medium	12401 (22.10)	1699 (22.10)	18001 (28.18)	2386 (28.18)
Large	26016 (20.92)	1618 (20.92)	25572 (20.48)	1698 (20.48)
All Farms	14159 (22.91)	1735 (22.91)	16934 (25.21)	2097 (25.21)

Table 5.3.3: Different Types of Costs Per Farm and Per Acre for Groundnut

(Value in '₹')

	Organic Farms		*Conventional Farms*	
Farm Size	*Per Farm*	*Per Acre*	*Per Farm*	*Per Acre*
Prime Cost				
Small	52236 (77.12)	15692 (77.12)	48274 (82.12)	16469 (82.12)
Medium	110860 (80.12)	14843 (80.12)	125281 (82.24)	16484 (82.24)
Large	250044 (80.92)	12662 (80.92)	305463 (81.48)	17812 (81.48)
All Farms	120944 (78.23)	13487 (78.23)	139103 (79.12)	16564 (79.12)
Operational Cost				
Small	40769 (60.19)	12247 (60.19)	36023 (61.28)	12290 (61.28)
Medium	80655 (58.29)	10799 (58.29)	90152 (59.18)	11862 (59.18)
Large	189202 (61.23)	9581 (61.23)	254215 (67.81)	14823 (67.81)
All Farms	96486 (62.41)	10759 (62.41)	117320 (66.73)	13970 (66.73)
Overhead Cost				
Small	18518 (27.34)	5563 (27.34)	17588 (29.92)	6000 (29.92)
Medium	35713 (25.81)	4782 (25.81)	44665 (29.32)	5877 (29.32)
Large	83801 (27.12)	4243 (27.12)	111043 (29.62)	6475 (29.62)0
All Farms	41618 (26.92)	4641 (26.92)	50142 (28.52)	5971 (28.52)
Paid-out Cost				
Small	45450 (67.10)	13654 (67.10)	40515 (68.92)	13822 (68.92)
Medium	88818 (64.19)	11892 (64.19)	114145 (74.93)	15019 (74.93)
Large	227055 (73.48)	11497 (73.48)	279745 (74.62)	16312 (74.62)
All Farms	108498 (70.18)	12099 (70.18)	130102 (74.00)	15492 (74.00)
Imputed Cost				
Small	19155 (28.28)	5754 (28.28)	17295 (29.42)	5900 (29.42)
Medium	29431 (21.27)	3940 (21.27)	38830 (25.49)	5109 (25.49)
Large	65508 (21.20)	3317 (21.20)	100509 (26.81)	5861 (26.81)
All Farms	36300 (23.48)	4048 (23.48)	46274 (26.32)	5510 (26.32)

As far as the redgram farming is considered, it is apparent from Table - 5.3.2 that the proportion of imputed cost to total cost per acre and per farm on conventional farms is higher by 2 per cent. With regard to different size groups of farms, conventional medium farms registered a higher difference by 6 per cent than the small farms (one per cent). Contrary to this, the conventional large farms registered lower proportion of imputed cost to total cost by marginally compared to their counter parts.

As far as the groundnut farming is considered, it is evident from Table - 5.3.3 that the conventional farms registered a higher proportion of imputed cost to total cost per acre and per farm than the conventional farms by 3 per cent. Farm size groups wise analysis reveals that the conventional large farms registered a much higher proportion by 6 per cent when compared to the other size groups of farms i.e., conventional small and conventional medium farms(1.14 per cent and 4.22 per cent respectively).

5.4 Returns From Farming

The per acre returns from cultivation in both the categories of farms are analysed by calculating the following concepts of returns viz., gross returns, farm business income, family labour income, farm investment income and net income. The details for all the selected three crops viz. Paddy, Groundnut and Redgram are presented in Tables 5.4.

5.4.1 Gross Income

Gross income per acre for all organic (paddy, redgram and groundnut) farmers is ₹ 30,221/-, ₹ 13646/- and ₹ 26335/- respectively and for conventional farmers it is ₹ 28,717/-, ₹ 12387/- and ₹ 24626/-respectively, which implies that the organic farmers are earning 5 per cent, 10 per cent and 7 per cent more income compared to the conventional farmers of paddy, redgram and groundnut. Except the large farmers of groundnut and small farmers of redgram, all the other groups of farmers from organic category are earning more income compared to their counterparts in the conventional category. Gross income per farm is also higher for organic category farms compared to the conventional category farms. The size group wise analysis also shows the same picture though with slight variations in the amounts. It can be concluded that the gross income per acre as well as per farm is more for organic category among all the sections of the farmers except small farmers of paddy and redgram.

5.4.2 Farm Business Income

Farm business income represents returns to the farmer's land, family labour, fixed capital and management. It is originated by deducting the Cost A_1 or A_2, as the case may be, from the gross returns. A perusal of Table 5.4 reveals that the farm business income per acre for organic farms is ₹ 16568/-, ₹ 7671/- and ₹ 10713/- for the three selected crops respectively and it is 16 per cent, 26 per cent and 48 per cent higher than the conventional farm holdings. The size group wise analysis exhibits a more or less similar picture with slight variation in percentages except the small farmers of redgram. The small farmers of organic redgram are getting lesser farm business income compared with the other groups of farmers and with other crops of farms also.

5.4.3 Family Labour Income

Family labour income gives the return to the family labour and management of the crop enterprise, which can be arrived at by deducting Cost – B_2 from gross returns. A keen observation of the Table - 5.4 also reveals that the family labour income per acre is positive for both the organic and conventional farmers and registered as ₹ 9,068/-, ₹ 6,421/- and ₹ 9,713/- for the selected three Organic crops respectively and ₹ 5,895/-, ₹ 4,417/- and ₹ 4,554/- for the selected three conventional crops. It can also be found that for all size groups of farmers of the selected crops in both organic and conventional category registered a profitable family labour income except for the small farmers of redgram. The small farmers of redgram on both organic and conventional category registered a positive family labour income, but the farmers of organic redgram are getting lesser amount of family labour income.

5.4.4 Farm Investment Income

Farm investment income represents income retained with the farmer for his investment and it comprises of the rental value of own land, interest on own fixed capital and returns to the management. The farm investment income per acre for organic farmers is reported as ₹ 16,981/-, ₹ 7,463/- and ₹ 10,757/- for the three selected crops respectively, while it is ₹ 14,231/-, ₹ 5,466/- and ₹ 5,105/- respectively for conventional category farmers, which reveals that organic farmers in the study

Table 5.4.1: Different Types of Returns from Cultivation of Paddy

(Value in '₹')

	Organic Farms		*Conventional Farms*	
Farm Size	*Per Farm*	*Per Acre*	*Per Farm*	*Per Acre*
Gross Returns				
Small	91485	28818	98648	28733
Medium	230059	30502	224912	29252
Large	543533	30424	466577	28353
All Farms	239854	30221	236085	28717
Farm Business Income				
Small	51699	16128	42189	12288
Medium	142032	18342	105805	13761
Large	307456	16693	238755	14509
All Farms	140797	16568	114232	13895
Family Labour Income				
Small	27889	8628	14722	4288
Medium	85464	10842	44294	5761
Large	173464	9193	107107	6509
All Farms	81272	9068	48464	5895
Farm Investment Income				
Small	54315	16952	73863	12966
Medium	143441	18342	100339	13705
Large	316223	17184	258194	14997
All Farms	144071	16981	197917	14231
Net Income				
Small	26375	8151	12576	3663
Medium	82153	10403	40460	5262
Large	168974	8942	102324	6218
All Farms	78392	8706	45051	5480

Table 5.4.2: Different Types of Returns from Cultivation of Redgram

(Value in' ")

	Organic Farms		*Conventional Farms*	
Farm Size	*Per Farm*	*Per Acre*	*Per Farm*	*Per Acre*
Gross Returns				
Small	39369	12721	49065	13905
Medium	98467	13494	90624	12013
Large	224639	13971	186191	12360
All Farms	111338	13646	100012	12387
Farm Business Income				
Small	23358	7548	26954	7639
Medium	54024	7403	38819	5146
Large	126219	7850	85444	5672
All Farms	62586	7671	45755	5667
Family Labour Income				
Small	19490	6298	22543	6389
Medium	44902	6153	29389	3896
Large	106121	6600	66615	4422
All Farms	52387	6421	35662	4417
Farm Investment Income				
Small	23563	7614	31250	7689
Medium	52266	7403	38417	4897
Large	122019	7589	120780	5451
All Farms	60890	7463	93027	5466
Net Income				
Small	18567	6000	21338	6047
Medium	42354	5804	26747	3545
Large	100279	6237	61327	4071
All Farms	49534	6071	32840	4067

Table 5.4.3: Different Types of Returns from Cultivation of Groundnut

(Value in' ")

	Organic Farms		Conventional Farms	
Farm Size	*Per Farm*	*Per Acre*	*Per Farm*	*Per Acre*
Gross Returns				
Small	103260	31022	70350	24000
Medium	205032	27454	183177	24102
Large	483088	24460	432075	25194
All Farms	236145	26335	206808	24626
Farm Business Income				
Small	41315	12413	16957	5785
Medium	79131	10597	44716	5884
Large	204813	10369	89761	5234
All Farms	96057	10713	46644	5554
Family Labour Income				
Small	37986	11413	14026	4785
Medium	71663	9597	37116	4884
Large	185063	9369	72611	4234
All Farms	87090	9713	38246	4554
Farm Investment Income				
Small	42729	12838	26948	5932
Medium	82047	10597	43309	5337
Large	199398	10095	208463	4728
All Farms	96448	10757	181378	5105
Net Income				
Small	35526	10674	11565	3945
Medium	66664	8928	30842	4058
Large	174086	8813	57182	3334
All Farms	81545	9095	30995	3691

area are getting 16 per cent, 27 per cent and 53 per cent higher farm investment incomes compared to their counterparts. It can also be found from the Table that the farm investment income for all the size-groups and for all the three crops is found to be higher for organic category except for the small farmers of redgram. The farm investment income of the small farmers of organic redgram is lower than small farmers of conventional redgram and the difference is registered as ₹ 75/- (0.97 per cent), which is a very negligible amount.

5.4.5 Net Income

Net income indicates the profit or loss from farm business. It is the residual of gross income after deducting total cost viz., Cost – C_2 from it. A close observation of the Table 5.4 reveals that the farmers of all size groups of the selected crops under both organic and conventional category are getting profits, but the profits earned by the organic farmers are higher by 37 per cent, 33 per cent and 59 per cent for the selected crops respectively. A more or less similar picture can be seen from the analysis of different size groups of farms on both the organic and conventional category of the selected crops except for the small farmers of redgram. The small farmers of organic redgram are getting lower profits or net incomes than their counterpart by ₹ 47/- (0.77 per cent), which is a very negligible amount.

On the basis of the preceding analysis, it can be concluded that farmers of both organic and conventional categories of all the crops are getting benefited with regard to the various standard concepts of returns employed and analysed in this Study. It can also be seen that the small size farmers of organic category of the redgram are getting lesser profits than their counterparts. Another important observation that can be made from the analysis is that the organic groundnut farmers of large farm size group are getting lesser profits than their counterparts.

SECTION – II

While Section – I has dealt with costs and returns of organic farming vis-à-vis conventional farming and concluded that the organic farmers are accruing higher income compared to the conventional farmers, an attempt is made in Section – II to analyse the experiences and perceptions of organic farmers to elicit information on (i) advantages or otherwise of organic farming, (ii) its impact on the village economy and social institutions of the village community, (iii) by whom they were motivated to go in for organic farming,(iv) the impact of organic farming on environment etc.

It is heartening to note that as many as 18 per cent have been adopting organic farming since 2001 and all of them have been continuing organic farming to date (see Table 5.5). Despite this fact about 15 per cent of them have switched over to organic farming only in 2005 and all the selected organic farmers have crossed the gestation period of three years and reaping the benefits of organic farming.

A glance at Table 5.6 reveals that electronic media has more impact on the switching over to organic farming as it is evident from the fact that it motivated around 21 per cent of the total sample farmers followed by village cooperative (19 per cent), print media (17 per cent), village leaders

(15 per cent), Agricultural Extension workers (15 per cent) and fellow farmers (13 per cent). Slight variations in the percentages, can be found at the crop level analysis also.

Table 5.5: Details of Experience in Organic Farming: Crop-wise

S. No.	*Year*	*Paddy*	*Redgram*	*Groundnut*	*Total*
1.	2001	32 (21.33)	17 (17.00)	13 (13.00)	62 (17.72)
2.	2002	33 (22.00)	19 (19.00)	18 (18.00)	70 (20.00)
3.	2003	24 (16.00)	25 (25.00)	22 (22.00)	71 (20.28)
4.	2004	36 (24.00)	26 (26.00)	31 (31.00)	93 (26.58)
5.	2005	25 (16.67)	13 (13.00)	16 (16.00)	54 (15.42)
	Total	150 (100.00)	100 (100.00)	100 (100.00)	350 (100.00)

Table 5.6 : Agency or Person Who Recommended Organic Farming: Crop-wise

S. No.	*Name of the Agency*	*Paddy*	*Redgram*	*Groundnut*	*Total*
1.	Extension Worker	24 (16.00)	16 (16.00)	12 (12.00)	52 (14.86)
2.	Fellow Farmer	21 (14.00)	9 (9.00)	14 (14.00)	44 (12.57)
3.	Village Leader	29 (19.33)	11 (11.00)	13 (13.00)	53 (15.14)
4.	Village Co-operative	12 (8.00)	29 (29.00)	26 (26.00)	67 (19.14)
5.	Print Media	28 (18.66)	13 (13.00)	18 (18.00)	59 (16.86)
6.	Electronic Media	36 (24.00)	22 (22.00)	17 (17.00)	75 (21.43)
	Total	150 (100.00)	100 (100.00)	100 (100.00)	350 (100.00)

It is distressing to note that out of the selected organic farmers none has reported that he has obtained certification, though as many as 62 per cent have reported that they have taken up organic farming in 2001. The sample farmers of the study area based on their experience in organic farming reported some advantages of organic farming which are correlated with the results of the earlier studies. Around 34 per cent of them reported that the fertility of soil is being increased because of organic farming. In addition, around 37 per cent of them reported that the cost of cultivation has come down due to non-usage of chemical fertilisers. Further around 15 per cent of them reported that the organic produce is good for health, while another 13 per cent of them have reported that they are getting higher and constant returns from organic farming (see Table 5.7).

Table 5.7: Advantages of Organic Farming

S. No.	*Advantage*	*Paddy*	*Redgram*	*Ground-nut*	*Total*
1.	Increases the Soil Fertility	45 (30.00)	33 (33.00)	41 (41.00)	119 (34.00)
2.	Lower Cost of Production	49 (32.67)	46 (46.00)	35 (35.00)	130 (37.14)
3.	Good for Health	35 (23.33)	4 (4.00)	15 (15.00)	54 (15.43)
4.	Yield is Constant Higher	21 (14.00)	17 (17.00)	9 (9.00)	47 (13.43)
	Total	150 (100.00)	100 (100.00)	100 (100.00)	350 (100.00)

With regard to the certification for organic produce, the sample farmers expressed that they are not getting certification for their organic produce. The reasons as expressed are, it is of highly expensive (66 per cent), followed by lack of information on the certification process (27 per cent) and small size of farm holdings (6.58 per cent). The crop wise results with regards to this aspect have been presented in Table 5.8.

Table 5.8: Reasons for not getting Certification for Organic Produce

S. No.	Reason	Paddy	Redgram	Groundnut	Total
1.	Highly expensive	95 (63.33)	71 (71.00)	65 (65.00)	231 (66.00)
2.	Lack of sufficient information	45 (30.00)	23 (23.00)	28 (28.00)	96 (27.42)
3.	Small of farm	10 (6.67)	6 (6.00)	7 (7.00)	23 (6.58)
	Total	150 (100.00)	100 (100.00)	100 (100.00)	350 (100.00)

When information was elicited as to other problems almost all of them reported that they have been facing problems in marketing their produce as their product lacks with certification. All of them reported difficulties in certification. The details can be observed in Table 5.9.

Table 5.9: Problems of Sample Farmers in Organic Farming

S. No.	Problem	Paddy	Redgram	Groundnut	Total
1.	Marketing the produce	143 (95.33)	92 (92.00)	97 (97.00)	332 (94.85)
2.	Difficulty in getting certification	150 (100.00)	100 (100.00)	100 (100.00)	350 (100.00)
3.	Lack of govt. support	150 (100.00)	100 (100.00)	100 (100.00)	350 (100.00)

Suggestions as made by the sample farmers to encourage organic farming have been presented in Table 5.10. It can be observed from the Table that all the sample farmers suggested that the organic farming will spread, if the govt. provides subsidies on organic inputs and support for getting certification and marketing the produce. In addition, they suggested that any technical support from the agricultural line department will also be quite helpful for them. As a whole, the farmers felt that it is in the hands of govt. to encourage the organic farming on a wider scale.

Table 5.10: Suggestions Provided by Organic Farmers for an Effective Spread of Organic Farming

S. No.	*Suggestions*	*Paddy*	*Redgram*	*Groundnut*	*Total*
1.	Subsidies of Organic Inputs (Vermi Compost)	140 (93.34)	85 (85.00)	89 (89.00)	314 (89.71)
2.	Govt. Support for Certification and Marketing	150 (100.00)	100 (100.00)	100 (100.00)	350 (100.00)
3.	Agriculture Line dept. for Technical Support	150 (100.00)	100 (100.00)	100 (100.00)	350 (100.00)

- The cost of paddy per farm and per acre on the basis of different cost concepts is found to be relatively higher on conventional farms compared to organic farms. The same phenomenon is discernible among different size groups of farms also.
- Further, it is to be noted that in case of organic paddy holdings, the proportions of different costs to Cost – $_{C2}$ and farms size are directly related, whereas in case of conventional holdings, the proportions of different costs to Cost – C_2 are inversely related.
- The cost of redgram per farm and per acre on the basis of different cost concepts is found to be relatively higher on conventional farms compared to organic farms. The same phenomenon is discernible among different size groups of farms also. Further, it is to be noted that in case of organic holdings, the proportions of different costs to Cost – C_2 are directly related, whereas in case of conventional holdings, the proportions of different cost s to Cost – C_2 are inversely related.
- The cost of groundnut per acre on the basis of different cost concepts is found to be relatively higher on conventional farms compared to organic farms. Except the small farm holdings, the same phenomenon is discernible among different size groups of farms also and the cost of cultivation for small farm holdings on organic farming is slightly higher. In case of organic holdings, the proportions of different costs to Cost – C_2 and farms size are inversely related, whereas in case of conventional holdings, these proportions are directly related.
- The farmers of all size groups of the selected crops under both organic and conventional category are getting profits, but the profits earned by the organic farmers are higher by 37 per cent, 33 per cent and 59

per cent for the selected crops respectively. A more or less similar picture can be seen from the analysis of different size groups of farms on both the organic and conventional category of the selected crops except for the small farmers of redgram. The small farmers of organic redgram are getting lower profits or net incomes than their counterpart by ₹ 47/- (0.77 per cent), which is a very negligible amount.

- When perceptions of organic farmers were elicited as to their experiences in organic farming, certification, problems they encounter with etc., it is heartening to note that as many as 18 per cent have been adopting organic farming since 2001 and all of them have been continuing it to date. Despite this fact, about 15 per cent of them have switched over to organic farming only in 2005 and all the selected organic farmers have crossed the gestation period of three years and have been reaping the benefits of organic farming.
- Electronic media has more impact on the switching over to organic farming , as it is evident from the fact that it motivated around 21 per cent of the total sample farmers, followed by village cooperative (19 per cent), print media (17 per cent), village leaders (15 per cent), agricultural extension workers (15 per cent) and fellow farmers (13 per cent). Slight variations in the percentages, can be found at the crop level analysis also.
- It is distressing to note that out of the selected organic farmers none has reported that he has obtained certification, though many of them have reported that they have taken organic farming as early as in 2001. The sample farmers of the study area based on their experience in organic farming reported some advantages of organic farming which are correlated with the results of the earlier studies. Around 34 per cent of them reported that the fertility of soil has increased. In addition, around 37 per cent of them reported that the cost of cultivation has come down due to non-usage of chemical fertilisers. Further, around 15 per cent of them reported that the organic produce is good for health, while another 13 per cent of them reported that they are getting higher and constant returns from organic farming.
- With regard to the certification for organic produce, they expressed, that certification is highly expensive (66 per cent), followed by lack of information on the certification process (27 per cent) and small size of farm holdings (7 per cent).
- When information was elicited as to other problems, almost all of them reported that they have been facing problems in marketing their produce as their product lacks certification.

- All the sample farmers suggested that the organic farming will spread if the government provides some subsidies on organic inputs and support them in getting certification and enable them to market their produce at remunerative prices. In addition, they suggested that any technical support from the agricultural line department will also be of quite help for them. As a whole, the farmers felt that it is in the hands of government to encourage the organic farming on a wider scale.
- On the basis of the preceding analysis, it can be concluded that farmers of both organic and conventional farmers are getting benefited with regard to the various standard concepts of returns employed and analysed in this Study. It can also be seen that the small farmers of organic redgram, are getting lower profits compared their counterparts. Another important observation that can be made from the analysis is that organic groundnut farmers of large farm size group are getting lower profits compared their counterparts. Based on these conclusions, it could not be generalised that the organic farmers are more efficient both technically and allocatively compared to the conventional farmers.

Note

Prasada Rao. B and Mohana Rao, L.K. (1986), "Studies in the Economics of Farm Management in the Command Area of Nagarjuna Sagar Irrigation Project, *DES, DA&C,Sponsored by Ministry of Agriculture,* Govt. of India, New Delhi.

Chapter 6

Economic Efficiency of Organic Farming *vis-a-vis* Conventional Farming

Analysis presented in Chapter – 5 has revealed that the intensity of input-use is higher in conventional farming compared to organic farming. Now an attempt is made in this Chapter to examine the economic efficiency of organic farming vis-à-vis conventional farming. This Chapter is divided into three sections Section – I deals with Stochastic Frontier Production Function (SFPF), Section – II deals with Data Envelopment Analysis (Computer) Programme (DEAP), while Section – III deals with Factors Determining Technical Efficiency.

SECTION – I

The measurement of the productive efficiency of a farm relative to other farms or to the "best practice" in an industry has long been of interest to agricultural economists. Much empirical work has centered on imperfect and partial measures of productivity, such as yield per hectare, output per unit of labour, etc. Farrell (1957)[1] suggested a method of measuring the technical efficiency of a firm in an industry by estimating the production function of firms which are "fully-efficient (i.e., a frontier production function).

Subsequently, some research studies have applied and extended Farrell's ideas. These studies may be broadly divided into two groups according to the method chosen to estimate the frontier production function, viz., mathematical programming versus econometric estimation. Debate still continues over which approach is the most appropriate method to use. The answer often depends upon the application considered. The mathematical programming approach to frontier estimation is usually termed Data Envelopment Analysis (DEA). Coelli(1995)[2] outlines the methodology, of estimation and the limitations of DEA.

Primary criticism of the DEA approach is that measurement errors can have larger influence upon the shape and positioning of the estimated frontier. Aigner, Lovell and Schmidt (1977)[3] and Meeusen and Van den Broeck (1977)[4] independently proposed the stochastic frontier production function to account for the presence of measurement production functions. Stochastic frontier production functions have two error terms, one to account for the existence of technical inefficiency of production and the other to account for factors such as measurement error in the output variable, and the combined effects of unobserved inputs on production. This favourable property of stochastic frontier production frontiers come with a price, namely, that the functional form of the production function and distributional assumptions of the two error terms must be explicitly specified. Bauer (1990)[5] and Greene (1993)[6] present comprehensive reviews of the econometric estimation of frontiers. Coelli (1995a)[7] also outlines models and application of Stochastic Frontier Production Functions.

In the agricultural economics literature, the stochastic frontier (econometric) approach has generally been preferred. This is probably associated with a number of factors. The assumption that all deviations from the frontier are associated with inefficiency, as assumed in DEA, is difficult to accept, given the inherent variability of agricultural production, due to weather, fires, pests, diseases, etc. Further, as many farms are small family-owned operations, the keeping of accurate records is not always a priority. Thus much available data on production are likely to be subject to measurement errors.

There have been many applications of frontier production functions to agricultural industries over the years. Battese (1992)[8] and Bravo Ureta and Pinherio (1993)[9] have provided a survey of applications in agricultural economics, the latter giving particular attention to applications in developing countries. Bravo-Ureta and Pinherio (1993)[10] also have drawn attention to those applications which attempt to investigate the relationship between technical efficiencies and various socio-economic variables, such as age and level of education of the farmer, farm size and utilization of extension services. The identification of those factors which influence the level of technical efficiencies of farmers is, undoubtedly, a valuable exercise. The information provided may be of significant use to policy makers attempting to raise the average level of farmer efficiency. Most of the applications which seek to explain the differences in technical efficiencies of farmers use a two-stage approach. The first stage involves the estimation of a stochastic frontier production function and the prediction of farm-level technical inefficiency effects (or technical efficiencies). In the second stage, these predicted technical inefficiency effects (or technical efficiencies) are related to farmer-specific factors using ordinary least squares regression. This approach appears to have been first used by Kalirajan (1981)[11] and has since been used by a larger number of agricultural economists. Prominent among them are the studies of Parikh and Shah (1994)[12], Kumbhakar, Ghosh and Mcguckin(1991)[13], Reifscheider and Stevenson(1991)[14], Huang and Lui(1994)[15] and Battese and Coelli (1995).[16] Most recently, in the context of Indian Agriculture, studies by C Ramasamy *et al.* (2003)[17] and D K Charyulu (2010)[18] have specified the stochastic frontiers and models for technical inefficiency effects and simultaneously estimate all parameters involved. This one-stage approach is less objectionable from a statistical point of view and is expected to lead to more efficient inference with respect to the parameters involved.

6.1 The Stochastic Frontier and Efficiency Model

The Stochastic Frontier Production Function (SFPF) Model which specify for the farming operations in a given farm category is as follows:

$$\ln(Y) = \beta_0 + \beta_1 \ln(Land) + \beta_2 \ln(HL) + \beta_3 \ln(TL) + \beta_4 (Seed) + \beta_5 \ln(OF) + \beta_6 \ln(OP) + \beta_7 \ln(CS) + U - V$$

ln represents the natural logarithm(i.e., to base e).

Y represents the total value of output (in '₹') from the crop which are grown.'

Land represents the total area of cropped land (in Acres)

HL represents the total quantity of human labour (family and hired labourers) measured in value terms (in '₹').

TL represents the total amount of bullock and machine labour (owned and hired) (in '₹').

Seed represents value of Seed (in '₹').

OF represents amount of organic fertilisers (in '₹') in case of organic farms and amount of chemical fertilizers in case of conventional farms.

OP represents the amount of organic pesticides (in '₹') in case of organic farms and amount on synthetic pesticides in case of conventional farms.

CS represents the amount on capital services (in '₹') which include depreciation on farm assets and interest on working capital.

The *Vs* are assumed to be independently and identically distributed random errors, having $N\ (0,\ \sigma^2)$ distribution; and

The *Us* are non-negative random variables, called technical efficiency effects, associated with the technical efficiency of production of the farmers involved.

The technical efficiencies under the above mentioned assumptions for organic farmers and conventional farmers and the parameters of the Model are estimated by the method of maximum likelihood, using the computer program, FRONTIER version 4.1 Coelli, (1992, 1994)[19]

The results of the estimated SFPF for the selected organic and conventional farms are presented in Tables – 6.1 to 6.3. In case of organic paddy farms, coefficients of all the inputs with few exceptions, have registered the expected signs with a priori economic logic (Table 6.1.1). Most of these coefficients are found to be statistically significant at probability levels ranging from one to 10 per cent. Only the coefficients associated with traction labour and organic pesticides in the medium farms function are negative. However, these coefficients are found to be not significant even ten per cent probability level. The results show that the per acre output in organic farms is positively related to coefficient of organic fertilizers, pesticides, human labour, seed, capital services incurred in production.

The significant value of γ indicates that the difference between observed output and actual output is not only due to factors that are beyond the farmer's control, but also due to some technical inefficiency. The value of γ (0.93) signifies that 93 per cent of the difference in observed and the frontier output is primarily due to factors which are under the control of the farms. The mean technical efficiency of organic farms is estimated as 93 per cent. This implies, using the existing inputs in an efficient manner, the organic farms can increase the output by seven per cent.

In case of conventional paddy farms (Table 6.1.2), coefficients of all the inputs with few exceptions have registered the positive signs. Most of these coefficients are statistically significant at probability levels ranging from one to ten per cent. Only the coefficients associated with synthetic pesticides in small farms and capital services in medium farms are found to be negative. However, both these coefficients are found to be not significant even at ten per cent probability level. The results show that the per acre output in conventional farms is positively related to fertilizers, pesticides, human labour, seed, capital services incurred in production.

The significant value of γ indicates that the difference between observed output and actual output is not only due to factors that are beyond the farmer's control, but also due to some technical inefficiency. The value of γ (0.95) signifies that 95 per cent of the difference in observed and the frontier output is primarily due to factors, which are under the control of the farms. The mean technical efficiency of organic farms is estimated as 89 per cent. This implies, using the existing inputs in an efficient manner, the organic farms can increase the output by 11 per cent.

A comparison of organic and conventional farm functions revealed that the elasticity coefficients of different variables in most of the functions are relatively higher in conventional farms than organic farms. In addition, the technical efficiency is found to be relatively higher in organic farms as compared to conventional farms.

Table 6.1.1: Estimated Frontier Equation for Paddy – Organic Farmers

Coefficients	*Small Farms*	*Medium Farms*	*Large Farms*	*All Farms*
Constant	1.33 (1.33)	4.11 (4.13)	9.39 (37.77)	0.051 (9.53)
Land	0.27 (0.38)	0.14 (-0.10)	0.67* (15.66)	0.37* (5.31)
Human Labour	0.49* (6.71)	0.48 (3.58)	0.25 (2.71)	0.33 (2.28)
Traction Labour	0.485* (7.43)	-0.015 (-0.023)	0.287** (2.05)	0.161*** (1.98)
Seed	0.184 (2.05)	0.32 (3.04)	0.12** (2.23)	0.21* (3.19)
Organic Fertilisers	0.18 (3.12)	0.33 (2.42)	0.10*** (2.99)	0.32* (5.67)
Organic Pesticides	0.027*** (1.93)	-0.019 (-0.065)	0.012 (2.12)	0.004* (3.69)
Capital Services	0.058 (2.34)	0.001 (2.09)	0.086*** (2.00)	0.019 (2.54)
σ^2	0.018* (5.37)	0.011* (3.20)	0.004* (2.66)	0.011* (2.89)
γ	0.89* (5.68)	0.92* (3.90)	0.95* (6.23)	0.93** (2.07)
Log-likelihood	-67.97	-94.38	-56.95	-161.99
Mean TE	0.90	0.91	0.94	0.93
N	55	66	29	150

Numbers in parentheses are the t-statistics*,** and *** denote that the coefficients are significant at 1 per cent, 5 per cent and 10 per cent respectively.

Table 6.1.2: Estimated Frontier Equation for Paddy – Conventional Farmers

Coefficients	*Small Farms*	*Medium Farms*	*Large Farms*	*All Farms*
Constant	8.32 (9.04)	23.81 (23.87)	3.1 (6.08)	8.05 (8.41)
Land	0.64* (4.37)	0.35 (0.41)	0.68* (3.33)	0.66* (5.37)
Human Labour	0.56* (2.72)	0.73* (8.31)	0.32** (2.52)	0.38* (2.94)
Traction Labour	0.45* (3.93)	0.53* (3.32)	0.21 (1.07)	0.28* (2.85)
Seed	0.37** (2.56)	0.51* (3.24)	0.17** (2.47)	0.21** (2.54)
Fertilisers	0.27** (2.67)	0.19** (2.32)	0.17* (2.88)	0.19* (2.96)
Pesticides	-0.03 (0.79)	0.22** (2.33)	0.19* (3.22)	0.13* (3.12)
Capital Services	0.10 (0.69)	-0.002 (0.007)	0.15** (2.13)	0.35* (3.10)
σ^2	0.024* (4.76)	0.01* (3.67)	0.04* (6.98)	0.026* (5.43)
γ	0.99* (14.32)	0.92* (6.96)	0.95* (7.83)	0.95* (6.83)
Log likelihood	38.75	47.85	58.48	97.99
Mean TE	0.88	0.92	0.96	0.89
N	39	36	25	100

Numbers in parentheses are the t-statistics*,** and *** denote that the coefficients are significant at 1 per cent, 5 per cent and 10 per cent respectively.

In case of organic redgram farms, all the coefficients of all the inputs except that associated with seed in small farms are significant and have expected positive signs (Table 6.2.1). The results show that the per acre output in organic farms is positively related to coefficient of organic fertilizers, pesticides, human labour, seed, capital services incurred in production. The elasticity coefficient associated with human labour is found to be relatively higher as compared to the elasticity coefficients associated with other variables.

The significant value of γ indicates that the difference between observed output and actual output is not only due to factors that are beyond the farmer's control, but also due to some technical inefficiency. The value of γ (0.87) signifies that 87 per cent of the difference in observed and the frontier output is primarily due to factors which are under the control of the farms. The mean technical efficiency of organic farms is estimated as 60 per cent. This implies, using the existing inputs in an efficient manner, the organic farms can increase the output by 40 per cent. Among different farms, the mean technical efficiency varies between 0.59 in small farms to 0.61 in medium farms.

In case of conventional redgram farms, coefficients of all the inputs except that associated with pesticides in medium and large farm functions are significant and have expected positive signs (Table 6.2.2). The results show that the per acre output in organic farms is positively related to coefficient of fertilizers, pesticides, human labour, seed, capital services incurred in production.

The significant value of γ indicates that the difference between observed output and actual output is not only due to factors that are beyond the farmer's control, but also due to some technical inefficiency. The value of γ (0.98) signifies that 98 per cent of the difference in observed and the frontier output is primarily due to factors, which are under the control of the farms. The mean technical efficiency of organic farms is estimated as 64 per cent. This implies, using the existing inputs in an efficient manner, the conventional farms can increase the output by 36 per cent and among different farms; it ranges from 60 per cent on medium farms to 68 per cent on large farms. A comparison of organic and conventional farms reveals that conventional farms are relatively more efficient than the organic farms in the production of redgram.

In case of organic groundnut farms, coefficients of all the inputs are significant and have expected positive signs (Table 6.3.1). The results show that the per acre output in organic farms is positively related to coefficient of organic fertilizers, pesticides, human labour, seed, capital services incurred in production.

The significant value of γ indicates that the difference between observed output and actual output is not only due to factors that are beyond the farmer's control, but also due to some technical inefficiency. The value of γ (0.72) signifies that 72 per cent of the difference in observed and the frontier output is primarily due to factors, which are under the control of the farms. The mean technical efficiency of organic farms is estimated as 81 per cent. This implies that, using the existing inputs in an efficient manner, the organic farms can increase the output by 19 per cent and it varies between 18 per cent on medium farms and 25 per cent on small farms.

In case of conventional groundnut farms, coefficients of most of the inputs are significant and have expected positive signs (Table 6.3.2). The results show that the per acre output in conventional farms is positively related to coefficient of fertilizers, pesticides, human labour, seed and capital services incurred in production.

Table 6.2.1: Estimated Frontier Equation for Redgram – Organic Farmers

Coefficients	*Small Farms*	*Medium Farms*	*Large Farms*	*All Farms*
Constant	8.04 (28.64)	17.41 (8.04)	18.9 (68.24)	21.84 (12.29)
Land	0.11* (3.04)	0.74* (8.38)	0.49* (2.32)	0.37* (6.28)
Human Labour	0.46** (2.31)	0.36** (2.42)	0.31* (3.41)	0.34* (3.53)
Traction Labour	0.29* (6.92)	0.26* (8.27)	0.39* (4.32)	0.23* (2.95)
Seed	0.31 (1.22)	0.22* (3.35)	0.35* (6.14)	0.17* (3.88)
Organic Fertilisers	0.19* (7.97)	0.21* (2.85)	0.23* (4.96)	0.21* (3.11)
Organic Pesticides	0.24* (9.82)	0.59* (4.89)	0.13* (8.37)	0.33** (2.45)
Capital Services	0.14* (9.73)	0.12* (2.91)	0.15* (5.91)	0.13* (3.54)
σ^2	0.017* (4.50)	0.004* (3.33)	0.011** (2.17)	0.012* (3.66)
γ	0.99* (9.27)	0.67* (4.18)	0.88* (5.25)	0.87* (7.82)
Log likelihood	56.57	53.86	35.71	119.99
Mean	0.59	0.61	0.60	0.60
N	38	34	28	100

Numbers in parentheses are the t-statistics*,** and *** denote that the coefficients are significant at 1 per cent, 5 per cent and 10 per cent respectively.

The significant value of γ indicates that the difference between observed output and actual output is not only due to factors that are beyond the farmer's control, but also due to some technical inefficiency. The value of γ (0.88) signifies that 88 per cent of the difference in observed and the frontier output is primarily due to factors, which are under the control of the farms. The mean technical efficiency of conventional farms is estimated as 76 per cent. This implies that, using the existing inputs in an efficient manner, the organic farms can increase the output by 24 per cent and it is ranging from 20 per cent on medium farms to 26 per cent on small farms. A comparison of organic and conventional farms reveals that organic farms are relatively more efficient than the conventional farms in the production of groundnut.

Table 6.2.2: Estimated Frontier Equation Redgram – Conventional Farmers

Coefficients	*Small Farms*	*Medium Farms*	*Large Farms*	*All Farms*
Constant	3.16* (3.14)	9.97* (10.91)	6.69* (5.59)	6.75* (4.87)
Land	0.51* (4.31)	0.48* (2.92)	0.45* (3.96)	0.50* (4.91)
Human Labour	0.35** (2.10)	0.24*** (2.08)	0.33* (3.99)	0.26* (2.78)
Traction Labour	0.055* (3.27)	0.34* (2.92)	0.22* (4.64)	0.19* (3.17)
Seed	0.40* (5.24)	0.42* (4.90)	0.34* (2.69)	0.37* (3.94)
Fertilisers	0.36*** (1.99)	0.24* (4.48)	0.37* (2.85)	0.28* (3.31)
Pesticides	0.30** (2.57)	-0.17 (1.20)	-0.19*** (2.00)	0.09*** (2.19)
Capital Services	0.21*** (2.08)	0.41*** (2.07)	0.33* (3.25)	0.25* (4.95)
σ^2	0.004* (3.23)	0.016* (5.02)	0.006* (2.99)	0.013* (3.97)
γ	0.99* (6.03)	0.98* (9.91)	0.99* (6.16)	0.98* (7.18)
Log likelihood	26.65	28.35	17.87	38.83
Mean TE	0.65	0.60	0.68	0.64
N	14	25	11	50

Numbers in parentheses are the t-statistics*,** and *** denote that the coefficients are significant at 1 per cent, 5 per cent and 10 per cent respectively.

Table 6.3.1: Estimated Frontier Equation for Groundnut – Organic Farmers

Coefficients	*Small Farms*	*Medium Farms*	*Large Farms*	*All Farms*
Constant	62.9 (11.34)	44.84 (16.01)	31.23 (43.35)	49.54 (10.47)
Land	0.21* (5.43)	0.17** (2.14)	0.26* (9.77)	0.19* (9.42)
Human Labour	-0.11 (-1.15)	-0.24 (1.37)	-0.15 (1.52)	-0.19 (-1.04)
Traction Labour	0.14* (5.44)	0.18* (5.35)	0.13* (7.03)	0.16* (7.02)
Seed	0.34* (6.87)	0.27* (6.27)	0.39* (8.06)	0.35* (5.97)
Organic Fertilisers	0.43* (6.48)	0.38* (5.93)	0.41* (6.91)	0.40* (7.32)
Organic Pesticides	0.13* (6.79)	0.18* (6.65)	0.09* (4.85)	0.14* (5.25)
Capital Services	0.16* (3.75)	0.04* (4.52)	0.15* (7.36)	0.07* (5.25)
σ^2	0.003* (4.08)	0.016* (6.59)	0.014** (2.14)	0.014* (4.37)
γ	0.89* (26.7)	0.79* (6.05)	0.83* (8.17)	0.82* (5.94)
Log -likelihood	68.99	54.9	28.58	102.39
Mean TE	0.75	0.82	0.79	0.81
N	35	41	24	100

Numbers in parentheses are the t-statistics*,** and *** denote that the coefficients are significant at 1 per cent, 5 per cent and 10 per cent respectively.

Table 6.3.2: Estimated Frontier Equation for Groundnut – Conventional Farmers

Coefficients	*Small Farms*	*Medium Farms*	*Large Farms*	*All Farms*
Constant	3.91 (9.98)	-2.96 (-4.5)	4.39 (8.72)	3.78 (9.82)
Land	0.16** (2.40)	0.24* (4.97)	0.43* (10.28)	0.32* (12.3)
Human Labour	0.16* (3.10)	0.28* (10.89)	0.31* (6.15)	0.27* (5.84)
Traction Labour	0.07** (2.49)	0.12* (2.85)	0.11* (4.01)	0.11* (7.82)
Seed	0.23* (4.73)	0.13* (2.97)	0.23* (8.54)	0.17* (6.73)
Fertilisers	0.19* (6.04)	0.22* (5.44)	0.17* (9.04)	0.19* (3.53)
Pesticides	0.10* (6.77)	0.18* (8.38)	0.21* (3.29)	0.19* (3.79)
Capital Services	0.06*** (2.14)	0.08* (4.76)	0.13* (5.95)	0.12* (4.08)
σ^2	0.018* (4.38)	0.015* (3.64)	0.002* (5.65)	0.034* (3.56)
γ	0.89* (9.54)	0.89* (2.02)	0.87* (6.77)	0.88* (8.34)
Log likelihood	19.89	27.75	25.5	45.38
Mean TE	0.74	0.80	0.79	0.76
N	16	22	12	50

Numbers in parentheses are the t-statistics*, ** and *** denote that the coefficients are significant at 1 per cent, 5 per cent and 10 per cent respectively.

Section - II

The objective of this Section is to carry out an empirical evaluation of the technical efficiency achieved by organic farms in comparison with conventional farms. The analysis has been carried out by utilizing the recently developed DEA Model (Lovell, 1993[20]; Coelli, 1996[21]; Kumaracharyulu and Subho, 2010[22]). Higher technical efficiency score of one sample farmer relative to his counterpart means that, on an average, the former lay closer to their specific production frontier than the sample counterpart does with their respective production frontier. Each observation consists of the gross value of production per acre as output (Y) and costs on five inputs, viz., human labour (X_1), traction labour (X_2), seed (X_3), fertiliser (X_4) and pesticides (X_5). Input oriented DEA Model is used and the analysis is carried out by using Data Envelopment Analysis (Computer) Programme (DEAP) 2.1 Coelli, (1996)[23].

6.2.1 Efficiency of Paddy Cultivation under CRS, VRS and SE

The frequency distribution, mean, maximum and minimum efficiencies under CRS (Constant Returns to Scale), VRS (Variable Returns to Scale) and SE (Scale Efficiency) models of the DEA approach for sample organic and conventional farms is presented in Table 6.4.1. The estimated mean CRS-TE, VRS-TE and Scale efficiencies for organic farms are 84 per cent, 86 per cent and 94 per cent while in conventional farms, they are 82 per cent, 86 per cent and 91 per cent respectively. Mean technical efficiency of CRS-TE, VRS-TE and SE models were higher in organic farms than conventional farms, relative to their specific frontiers. This implies that organic farms operate close to their specific frontier than conventional farms.

In terms of technical efficiency, 69.33 per cent, 75.33 per cent and 98 per cent of organic farms are more than 75 per cent efficiency under the CRS-TE, VRS-TE and SE models. Similarly, the same proportions are worked out to 64 per cent, 78 per cent and 89 per cent respectively in conventional farms.

Table 6.4.1: Frequency Distribution of Efficiency of Cultivation under CRS, VRS and SE – Paddy Farms

Efficiency %	*Organic Farms (N = 150)*			*Conventional Farms (N = 100)*		
	CRS-TE	*VRS-TE*	*SE*	*CRS-TE*	*VRS-TE*	*SE*
>25%	0.00	0.00	0.00	0.00	0.00	0.00
26-50%	0.67	0.00	0.00	0.00	0.00	0.00
51-75%	30.00	24.67	2.00	36.00	32.00	11.00
76-100%	69.33	75.33	98.00	64.00	78.00	89.00
Max(%)	100.00	100.00	100.00	100.00	100.00	100.00
Min(%)	49.00	52.00	68.00	61.00	64.00	66.00
Mean(%)	84.00	86.00	94.00	82.00	86.00	91.00

6.2.2 Efficiency of Redgram Cultivation under CRS, VRS and SE

It is evident from Table 6.4.2, that the estimated mean CRS-TE, VRS-TE and SE efficiencies for organic farms are 58 per cent, 60 per cent and 68 per cent, while in conventional farms they are 61 per cent, 64 per cent and 72 per cent respectively. Mean technical efficiencies of CRS-TE, VRS-TE and SE models are higher in conventional farms than organic farms, relative to their specific frontiers. This implies that conventional farms operate close to their specific frontier than organic farms.

In terms of technical efficiency, 44 per cent, 34 per cent and 41 per cent of organic farms are more than 75 per cent efficiency under the CRS-TE, VRS-TE and SE models. Similarly, in case of conventional farms, these efficiencies are 40 per cent, 34 per cent and 32 per cent respectively.

Table 6.4.2: Frequency Distribution of Efficiency of Cultivation under CRS, VRS and SE – Redgram Farms

Efficiency %	*Organic Farms (N = 100)*			*Conventional Farms (N = 50)*		
	CRS-TE	*VRS-TE*	*SE*	*CRS-TE*	*VRS-TE*	*SE*
>25%	0.00	0.00	0.00	0.00	0.00	0.00
26-50%	10.00	0.00	0.00	0.00	0.00	0.00
51-75%	46.00	66.00	59.00	60.00	66.00	68.00
76-100%	44.00	34.00	41.00	40.00	34.00	32.00
Max(%)	100.00	100.00	100.00	100.00	100.00	100.00
Min(%)	48.00	51.00	62.00	53.00	58.00	64.00
Mean(%)	58.00	60.00	68.00	61.00	64.00	72.00

6.2.3 Efficiency of Groundnut Cultivation under CRS, VRS and SE

It is evident from Table 6.4.3, that the estimated mean CRS-TE, VRS-TE and SE efficiencies for organic farms are 78 per cent, 81 per cent and 83 per cent while in conventional farms, these are 75 per cent, 75 per cent and 79 per cent respectively. Mean technical efficiencies of CRS-TE, VRS-TE and SE models were higher in organic farms than conventional farms, relative to their specific frontiers. This implies that organic farms operate to close their specific frontier than conventional farms.

In terms of technical efficiency, 55 per cent, 68 per cent and 69 per cent of organic farms and 46 per cent, 48 per cent and 60 per cent of conventional farms are more than 75 per cent efficiency under the CRS-TE, VRS-TE and SE models.

Table 6.4.3: Frequency Distribution of Efficiency of Cultivation under CRS, VRS and SE – Groundnut Farms

Efficiency %	*Organic Farms (N = 100)*			*Conventional Farms (N = 50)*		
	CRS-TE	*VRS-TE*	*SE*	*CRS-TE*	*VRS-TE*	*SE*
>25%	0.00	0.00	0.00	0.00	0.00	0.00
26-50%	0.00	0.00	0.00	0.00	0.00	0.00
51-75%	45.00	32.00	31.00	54.00	52.00	40.00
76-100%	55.00	68.00	69.00	46.00	48.00	60.00
Max(%)	100.00	100.00	100.00	100.00	100.00	100.00
Min(%)	56.00	57.00	61.00	55.00	57.00	62.00
Mean(%)	78.00	81.00	83.00	75.00	75.00	79.00

6.2.4 EFFICIENCY OF PADDY CULTIVATION (TE, AE AND EE) UNDER CRS:

The frequency distribution, mean, maximum and minimum of TE (Technical Efficiency), AE (Allocative efficiency) and EE (Economic Efficiency) under CRS (Constant Returns to Scale) model of DEA approach for sample organic and conventional farms is presented in Table 6.5.1. The estimated mean of TE,AE and EE for organic farms are 88 per cent, 90 per cent and 79 per cent respectively while for conventional farms, they are 88 per cent, 82 per cent and 74 per cent respectively.

In terms of technical efficiency, 93.33 per cent, 94 per cent and 74 per cent of organic farms are more than 75 per cent efficiencies of TE, AE and EE under CRS model. Similarly in case of conventional farms, 92 per cent,

75 per cent and 49 per cent have achieved more than 75 per cent efficiencies of TE, AE and EE respectively. The analysis shows that organic farms appear to be relatively more efficient than conventional farms under these three approaches.

Table 6.5.1: Frequency Distribution of Efficiency of Cultivation (TE,AE and EE) under CRS – Paddy Farms

Efficiency %	*Organic Farms (N = 150)*			*Conventional Farms (N = 100)*		
	CRS			*CRS*		
	TE	*AE*	*EE*	*TE*	*AE*	*EE*
>25%	0.00	0.67	0.67	0.00	0.00	0.00
26-50%	0.00	0.00	0.00	0.00	0.00	6.00
51-75%	6.67	5.33	25.33	8.00	25.00	45.00
76-100%	93.33	94.00	74.00	92.00	75.00	49.00
Max(%)	100.00	100.00	100.00	100.00	100.00	100.00
Min(%)	66.00	9.00	9.00	57.00	53.00	38.00
Mean(%)	88.00	90.00	79.00	88.00	82.00	74.00

6.2.5 EFFICIENCY OF REDGRAM CULTIVATION (TE, AE AND EE) UNDER CRS

It is evident from Table 6.5.2, that the estimated mean of TE, AE and EE under CRS model for organic farms are 61 per cent, 64 per cent and 45 per cent while, the same for conventional farms are 68 per cent, 74 per cent and 57 per cent respectively.

In terms of technical efficiency, 37 per cent, 28 per cent and five per cent of organic farms have attained more than 75 per cent efficiencies of TE, AE and EE under CRS model. While the same for conventional farms are 44 per cent, 38 per cent and 30 per cent respectively. The analysis reveals that conventional farms are more efficient compared to organic farms.

Table 6.5.2: Frequency Distribution of Efficiency of Cultivation (TE,AE and EE) under CRS – Redgram Farms

Efficiency %	*Organic Farms (N = 100)*			*Conventional Farms (N = 50)*		
	CRS			CRS		
	TE	*AE*	*EE*	*TE*	*AE*	*EE*
>25%	0.00	0.00	0.00	0.00	0.00	0.00
26-50%	10.00	18.00	56.00	0.00	0.00	10.00
51-75%	53.00	54.00	39.00	56.00	62.00	60.00
76-100%	37.00	28.00	5.00	44.00	38.00	30.00
Max(%)	100.00	100.00	100.00	100.00	100.00	100.00
Min(%)	47.00	31.00	27.00	57.00	51.00	37.00
Mean(%)	61.00	64.00	45.00	68.00	74.00	57.00

6.2.6 EFFICIENCY OF GROUNDNUT CULTIVATION (TE, AE AND EE) UNDER CRS:

It can be observed from Table 6.5.3, that the estimated mean efficiencies of TE, AE and EE under CRS model for organic farms are 82 per cent, 91 per cent and 78 per cent, while in conventional farms these are 75 per cent, 83 per cent and 66 per cent respectively.

In terms of technical efficiencies, 56 per cent, 76 per cent and 70 per cent of organic farms are more than 75 per cent efficiencies of TE, AE and EE under CRS model. Similarly in case of conventional farms these efficiencies are 26 per cent, 62 per cent and 24 per cent. It indicates that organic farms are more efficient as compared to conventional farms.

Table 6.5.3: Frequency Distribution of Efficiency of Cultivation (TE,AE and EE) under CRS – Groundnut Farms

Efficiency %	*Organic Farms (N = 100)*			*Conventional Farms (N = 50)*		
	CRS			*CRS*		
	TE	*AE*	*EE*	*TE*	*AE*	*EE*
>25%	0.00	0.00	0.00	0.00	0.00	0.00
26-50%	3.00	3.00	4.00	2.00	0.00	12.00
51-75%	41.00	21.00	26.00	72.00	38.00	64.00
76-100%	56.00	76.00	70.00	26.00	62.00	24.00
Max(%)	100.00	100.00	100.00	100.00	100.00	100.00
Min(%)	41.00	48.00	41.00	50.00	64.00	38.00
Mean(%)	82.00	91.00	78.00	75.00	83.00	66.00

6.2.7 EFFICIENCY OF PADDY CULTIVATION (TE, AE AND EE) UNDER VRS:

The frequency distribution of mean, maximum and minimum of TE (Technical Efficiency), AE (Allocative efficiency) and EE (Economic Efficiency) under VRS (Variable Returns to Scale) model of DEA approach for sample organic and conventional paddy farms is presented in Table 6.6.1. The estimated means of TE, AE and EE efficiencies for organic farms are 92 per cent, 90 per cent and 84 per cent while in conventional farms the same are 91 per cent, 84 per cent and 79 per cent respectively.

In terms of technical efficiency, 93.33 per cent, 96 per cent and 91 per cent of organic farms have attained more than 75 per cent efficiencies of TE, AE and EE under VRS model. Similarly 99 per cent, 76 per cent and 61 per cent of conventional farms have attained more than 75 per cent efficiencies under these three categories respectively. The results further indicate that the efficiencies under TE, AE and EE situations are respectively higher in organic farms as compared to conventional farms.

Table 6.6.1: Frequency Distribution of Efficiency of Cultivation (TE,AE and EE) under VRS – Paddy Farms

Efficiency %	*Organic Farms (N = 150)*			*Conventional Farms (N = 100)*		
	VRS			*VRS*		
	TE	*AE*	*EE*	*TE*	*AE*	*EE*
>25%	0.00	0.67	0.67	0.00	0.00	0.00
26-50%	0.00	0.00	0.00	0.00	0.00	2.00
51-75%	0.67	4.00	8.67	1.00	24.00	37.00
76-100%	99.33	96.00	91.33	99.00	76.00	61.00
Max(%)	100.00	100.00	100.00	100.00	100.00	100.00
Min(%)	69.00	9.00	9.00	74.00	56.00	49.00
Mean(%)	92.00	90.00	84.00	91.00	84.00	79.00

6.2.8 EFFICIENCY OF REDGRAM CULTIVATION (TE, AE AND EE) UNDER VRS

It can be found Table 6.6.2, that the estimated mean of TE, AE and EE efficiencies under VRS model for organic redgram farms are 56 per cent, 59 per cent and 57 per cent respectively, while in conventional farms the same are 59 per cent, 63 per cent and 60 per cent respectively.

In terms of technical efficiency, 30 per cent, 22 per cent and 17 per cent of organic farms are more than 75 per cent efficiencies of TE, AE and EE under VRS model. On the other hand, in case of conventional farms, the proportion of farms with more than 75 per cent efficiencies under TE, AE and EE situations are 24 per cent, 32 per cent and 22 per cent respectively. Further, 61 per cent, 60 per cent and 68 per cent of the organic farms under TE, AE and EE situations have attained efficiency in the range of 51-75 per cent, while the same in the case of conventional farms are 76 per cent and 68 per cent and 78 per cent respectively. The analysis reveals that conventional farms are more efficient compared to organic farms.

Table 6.6.2: Frequency Distribution of Efficiency of Cultivation (TE,AE and EE) under VRS –Redgram Farms

Efficiency %	*Organic Farms (N = 100)*			*Conventional Farms (N = 50)*		
	VRS			*VRS*		
	TE	*AE*	*EE*	*TE*	*AE*	*EE*
>25%	0.00	0.00	0.00	0.00	0.00	0.00
26-50%	9.00	18.00	15.00	0.00	0.00	0.00
51-75%	61.00	60.00	68.00	76.00	68.00	78.00
76-100%	30.00	22.00	17.00	24.00	32.00	22.00
Max(%)	100.00	100.00	100.00	100.00	100.00	100.00
Min(%)	38.00	42.00	39.00	52.00	56.00	51.00
Mean(%)	56.00	59.00	57.00	59.00	63.00	60.00

6.2.9 EFFICIENCY OF GROUNDNUT CULTIVATION (TE, AE AND EE) UNDER VRS:

It is evident from the Table 6.6.3, that the estimated means of TE, AE and EE efficiencies under VRS model for organic groundnut farms are 75 per cent, 94per cent and 74 per cent, while the same in conventional farms are 73 per cent, 82 per cent and 65 per cent respectively.

In terms of technical efficiencies, 24 per cent, 60 per cent and 22 per cent of organic farms have attained more than 75 per cent efficiencies of TE, AE and EE under VRS model. Similarly 32 per cent, 58 per cent and 24 per cent of conventional farms have attained efficiencies of 75 per cent and more under the same situations. It indicates that organic farms are more efficient compared to conventional farms.

Table 6.6.3: Frequency Distribution of Efficiency of Cultivation (TE,AE and EE) under VRS –Groundnut Farms

Efficiency %	*Organic Farms (N = 100)*			*Conventional Farms (N = 50)*		
	VRS			*VRS*		
	TE	*AE*	*EE*	*TE*	*AE*	*EE*
>25%	0.00	0.00	0.00	0.00	0.00	0.00
26-50%	2.00	0.00	2.00	0.00	0.00	10.00
51-75%	74.00	40.00	76.00	68.00	42.00	66.00
76-100%	24.00	60.00	22.00	32.00	58.00	24.00
Max(%)	100.00	100.00	100.00	100.00	100.00	100.00
Min(%)	48.00	67.00	48.00	51.00	68.00	42.00
Mean(%)	75.00	94.00	74.00	73.00	82.00	65.00

SECTION – III

In the earlier Section the economic efficiency of farmers under various conditions have been estimated and analysed. The results indicated that the technical efficiency in the use of resources under various conditions is relatively higher in case of paddy and significantly lower in case of redgram. Generally, the technical efficiency is influenced by several factors – technical, socio-economic and demographic factors. Kalirajan and Shand (1994)[24] have aptly pointed out that the technical efficiency is influenced by the technical knowledge and understanding as well as by socio-economic environment under which the farmers make decisions. Keeping this in view an attempt has been made in this Section to examine the factors that determining technical efficiency of organic and conventional farming for the three selected crops.

The following multiple regression model has been employed to analyse the factors determining technical efficiency of farmers:

$$TE_i = \beta_0 + \beta_1 X_1 + \beta_2 X_2 + \beta_3 X_3 + \beta_4 X_4 + \beta_5 D_1 + \beta_6 D_2 + e_t$$

TE_i = Technical Efficiency of i-th farmer

X_1 = Age of the farmer in years

X_2 = Years of schooling of the farmer

X_3 = Distance to the market (in Kilometers)

X_4 = Experience of the farmer in farming years (for organic farmers)

D_1 = 1, if belongs to small farmers
0, if otherwise

D_2 = 1, if belongs to small farmers
0, if otherwise

e = error term

$\hat{a}_1$ = regression coefficients to be estimated (i = 0, 1........6)

6.3.1 METHOD OF ESTIMATION:

The technical efficiency parameter lies in between 0 and 1. In such a type of situation a limited dependent variable estimation technique, like Tobit model, is often used by researchers. The underlying assumption of the Tobit model is that the dependent variable is censored and there are some underlying latent variables which are not observed. In this Study, all the values of TE_i are observed and there are no latent values. In addition, the results indicated that none of the technical efficiency scores has taken the value of zero. As aptly pointed out by Greene (2000),[25] if there is no observation with $TE_i = 0$, the Tobit approach is equalent to the OLS approach. Thus, in the present Study OLS method of estimation is employed to determine the factors influencing technical efficiencies.

6.3.2 VARIABLES USED

Age (X_1)

This variable refers to the age of the farmer in years. Generally, those farmers in the younger age groups are inclined to adopt innovative practices and thereby lead to an improvement in technical efficiency. Thus, an inverse relationship is hypothesized between age and technical efficiency.

Education (X_2)

An educated farmer has a relatively higher access to knowledge in modern practices in agriculture, technical knowhow, cultural practices etc., which may result in an improvement in technical efficiency. So, a positive relationship is hypothesized between the level of education of the farmer and technical efficiency.

Distance to Market (X_3)

A farmer living nearer to a market terminal is in a position to employ the resources on time and thereby may improve the technical efficiency. In addition, he may have more access to knowledge sources, as market is a place of not only providing inputs but also may be a platform for exchange of knowledge among different farmers or between input dealers and farmers or between farmers and technical persons of the input supply companies etc. Hence, an inverse relationship is hypothesized between distance to market and technical efficiency.

Experience (X_4)

This variable is used only in the organic farming models and is measured as the years of experience in organic farming. A positive relationship is hypothesized between experience and technical efficiency.

Farm Size (D_1, D_2)

An inverse relationship is hypothesized between farm size and technical efficiency. Hence, positive coefficients are expected for D_1 and D_2

6.3.3 Results

The results of the estimated regression functions are presented in Table 6.7. The Table reveals that the coefficient of the multiple determination is significant at one per cent probability level. In case of organic farms the explanatory power of the model (R^2) varies between 48 per cent in redgram to 58 per cent in paddy. This implies that all explanatory variables together are explaining 58 per cent, 48 per cent and 52 per cent of the variation in technical efficiency in case of paddy, redgram and groundnut respectively (Table 6.7.1). All the coefficients, with the exception of farm size dummies, have registered the expected signs and found to be significant at probability levels ranging from 1 to 10 per cent.

In case of conventional farms, the explanatory power of the model (R^2) ranges from 42 per cent in redgram to 55 per cent in paddy (Table 6.7.2) and the coefficients of multiple determination in all the three models is found to be significant at one per cent probability level. Most of the coefficients in different functions have registered the expected signs and found to be significantly different from zero at probability levels ranging from 1 to 10 per cent. In all the functions the coefficients associated with size dummies turned to be negative and found to be significant at probability levels ranging from 1 to 10 per cent. Further, the coefficient associated with the age of farmer for red gram, though positive, against the exceptions, however, found to be not significant even at 10 per cent probability level.

Table 6.7.1: Results of the Estimated Regression Equation – Organic Farmers

Coefficients	Paddy	Redgram	Groundnut
Constant	0.916	0.714	0.965
X_1	- 0.004*** (1.95)	- 0.003* (3.30)	- 0.002* (2.83)
X_2	0.017* (2.98)	0.009** (2.29)	0.006** (2.64)
X_3	- 0.008* (3.56)	- 0.004*** (1.92)	- 0.003*** (1.97)
X_4	0.135* (3.97)	0.019* (2.72)	0.058* (2.85)
D_1	- 0.011** (2.11)	- 0.022*** (1.99)	- 0.10* (4.35)
D_2	- 0.008** (2.67)	- 0.019* (3.01)	- 0.039** (2.25)
R^2	0.58	0.48	0.52
F - Value	13.46	11.49	12.16
N	150	100	100

Numbers in parentheses are the t-statistics*,** and *** denote that the coefficients are significant at 1 per cent, 5 per cent and 10 per cent respectively.

The negative sign associated with farm size dummies needs explanation. Earlier, prior to the ushering of Green Revolution in India, an inverse relationship exists between farm size on one hand and productivity and returns to scale on other hand. Basing on this an inverse relationship has been hypothesized between farm size and technical efficiency (Bagi 1981[26] and Sekar *et al* 1994[27]). However, the post-Green Revolution studies provide an inconclusive evidence on the inverse relationship between farm size and productivity and some studies indicated that productivity differences are size-neutral. In the present context, the negative sign of the farm size dummies indicate that big farms are more technically efficient than the medium and small farmers. Better access to credit, marketing facilities and agricultural extension services might have contributed to their higher efficiency.

Table 6.7.2: Results of the Estimated Regression Equation – Conventional Farmers

Coefficients	*Paddy*	*Redgram*	*Groundnut*
Constant	0.869	0.823	0.784
X_1	- 0.001** (2.17)	0.001 (1.67)	-0.0008*** (1.93)
X_2	0.005** (2.46)	0.003** (2.58)	0.004** (2.47)
X_3	- 0.006* (2.94)	- 0.018* (3.44)	- 0.003** (2.52)
D_1	- 0.045** (2.27)	- 0.051** (2.48)	- 0.032 (1.71)
D_2	- 0.026*** (1.91)	- 0.227* (5.02)	- 0.014*** (2.09)
R^2	0.55	0.42	0.47
F - Value	15.78	10.63	11.64
N	100	50	50

Numbers in parentheses are the t-statistics*,** and *** denote that the coefficients are significant at 1 per cent, 5 per cent and 10 per cent respectively.

Summary

- Results of the Stochastic Frontier Production Function Model indicated that technical efficiency is relatively higher on organic paddy farms compared to conventional paddy farms, while conventional redgram farmers are more efficient compared to their counterparts and organic groundnut farmers are relatively more efficient than their counterparts.
- An analysis of CRS-TE, VRS-TE and SE Model concluded that both organic paddy and groundnut farms operate close to their specific frontiers than conventional farms, while conventional redgram farms operate close to their specific frontier than organic redgram farms.
- An analysis of TE, AE and EE - CRS Model concluded that organic paddy and groundnut farms are more efficient compared to conventional paddy and groundnut farms, while conventional redgram farms are more efficient compared to organic redgram farms.
- Similarly an analysis of TE, AE and EE - VRS Model concluded that organic paddy and groundnut farms are more efficient compared to

conventional paddy and groundnut farms, while conventional redgram farms are more efficient as compared to organic redgram farms.

- Further an analysis of Factors Determining Technical Efficiency Model concluded that age of the farmer, education, distance to market and experience of the farmer appear to be predominant variables determining technical efficiency and large farms are found to be more efficient than medium and small farms.

Table 6.8: Summary of Results of Technical Efficiency of Organic and Conventional Farmers

Crop	*Organic*			*Conventional*		
Model - I	*CRS-TE*	*VRS-TE*	*SE*	*CRS-TE*	*VRS-TE*	*SE*
Paddy	✓	✓	✓	×	×	×
Redgram	×	×	×	✓	✓	✓
Groundnut	✓	✓	✓	×	×	×
	CRS			CRS		
Model - II	TE	AE	EE	TE	AE	EE
Paddy	*	✓	✓	*	×	×
Redgram	×	×	×	✓	✓	✓
Groundnut	✓	✓	✓	×	×	×
	VRS			VRS		
Model - III	TE	AE	EE	TE	AE	EE
Paddy	✓	✓	✓	×	×	×
Redgram	×	×	×	✓	✓	✓
Groundnut	✓	✓	✓	×	×	×

Note: ✓ - More Efficient, * - Equally Efficient, × - Less Efficient.
Source: Means of Efficiency Tables

Notes

1 Farrell, M.J. (1957), "The Measurement of Productive Efficiency", *Journal of the Royal Statistical Society,* A CXX, Part 3, 253-290.

2 Coelli, T.J. (1995), "Estimators and Hypothesis Tests for a Stochastic: A Monte Carlo Analysis", *Journal of Productivity Analysis*, 6, 247-268.

3 Aigner, D.J., Lovell, C.A.K. and Schmidt, P. (1977), Formulation and Estimation of Stochastic Frontier Production Function Models", *Journal of Econometrics*, 6 , 21-37.

4 Meeusen, W. and J. van den Broeck (1977), "Efficiency Estimation from Cobb-Douglas Production Functions with Composed Error", *International Economic Review*, 18, 435-444.

5 Bauer, P.W.(1990), "Rewcent Developments in the Econometric Estimation of

Frontiers", Journal of Econometrics, 46, 39-56.

6 Greene, W.H. (1993), "The Econometric Approach to Efficiency Analysis", in Fried, H.O., C.A.K. Lovell and S.S. Schmidt(eds), *The Measurement of Productive Efficiency: Techniques and Applications*, Oxford University Press, New York, pp. 68-119.

7 Coelli, T.J. (1995a), "Recent Developments in Frontier Estimation and Efficiency Measurement", *Australian Journal of Agriculture Economics*, 39, 219-245.

8 Battese, G.E.(1992), "Frontier Production Functions and Technical Efficiency: A Survey of Empirical Applications in Agricultural Economics", *Agricultural Economics*, 7, 185-208.

9 Bravo-Ureta, B.E. and A.E.Pinheiro (1993), "Efficiency Analysis of Developing Country Agriculture: A Review of the Frontier Function Literature", *Agricultural and Resource Economic Review*, 22, 88-101.

10 *Ibid.*

11 Kalirajan, K.P. (1981), "An Econometric Analysis of Yield Variability in Paddy Production", *Canadian Journal of Agricultural Economics*, 29,283-294.

12 Parikh, A. and K.Shah(1994), " Measurement of Technical efficiency in the North West Frontier Provience of Pakistan", *Journal of Agricultural Economics*, 45, 132-138.

13 Kumbhakar, S.C., S.Ghosh and J.T.McGuckin (1991), "A Generalised Production Frontier Approach for Estimating Determinants of Inefficiency in U.S.Dairy Farms", *Journal of Business and Economic Statistics*, 9,279-286.

14 Reifschneider, D. and R. Stevenson (1991), "Systematic Departures from the Frontier: A Framework for the Analysis of Firm Inefficiency", *International Economic Review*, 32, 715-723.

15 Huang, C.J. and J-T. Liu (1994), "Estimation of a Non-neutral Stochastic Frontier Production Function", *Journal of Productivity Analysis*, 4, 171-180.

16 Coelli, T.J(1995b), "A Monte Carlo Analysis of the Stochastic Frontier Production Function", Journal of Productivity Analysis, 6, 247-268.

17 Ramsamy, C. *et al.*, (2003) "Hybrid Rice in Tamil Nadu Evaluation of Farmers' Experience" *Economic and political Weekly*, June 21 2003, pp.2509-2512.

18 Kurma Charyulu D and Subho Biswas (2010), "Economics and Efficiency of Organic Farming vis-à-vis Conventional Farming in India" *Working Paper No. 2010-04-03*, CMA, IIM Ahmadabad, April 2010

19 Coelli, T.J.(1994) A Guide to FRONTIER *Version 4.1: A Computer Programme for Stochastic Frontier Production and Cost Function Estimation*, mimeo, Department of Econometrics, University of New England, Armidale, pp.32.

20 Lovell, C.A.K. (1993), "Production Frontiers and Productive Efficiency", in Fried, H.O., C.A.K. Lovell and S.S. Schmidt (Eds), *The Measurement of Productive Efficiency*, Oxford University Press, New York, 3-67.

21 Coelli, T.J., (1996) A Guide to DEAP 2.1: A Data Envelopment Analysis Computer Program, CEPA *working paper No.8/96*, ISBN 1863894969, Department of Econometrics, University of New England, Pp: 1-49

22 Kurma Charyulu, D and Subho Biswas (2010), "Efficiency of Organic Input Units under NPOF Scheme in India" *Working Paper No. 2010-04-01*, CMA, IIM Ahmadabad, April 2010.

23 *op.cit.*

24 Kalirajan, K.P and R.T. Shand(1994), "Economics in Disequilibrium: An Approach from Frontier", *Macmillan India Limited*, New Delhi.

25 Greene, W. (2000), "Econometric Analysis", 4th Edition, *Prentice Hall*, Upper Saddle River, New Jersey.

26 Bagi, S.F. (1981), "Relationship between Farm Size and Economic Efficiency: An Analysis of Farm Level Data from Haryana (India)", *Canadian Journal of Agricultural Economics* 29:317-326.

27 Sekar, C., C. Ramasamy and S.Senthilnathan (1994), "Size Productivity Relations in Paddy Farms of Tamil Naidu", *Agricultural Situation in India* 48: 859-863.

References

Aigner, D.J., Lovell, C.A.K. and Schmidt, P. (1977), Formulation and Estimation of Stochastic Frontier Production Function Models", *Journal of Econometrics*, 6, 21-37.

Bagi, S.F. (1981), "Relationship between Farm Size and Economic Efficiency: An Analysis of Farm Level Data from Haryana (India)", *Canadian Journal of Agricultural Economics* 29:317-326.

Battese, G.E.(1992), "Frontier Production Functions and Technical Efficiency: A Survey of Empirical Applications in Agricultural Economics", *Agricultural Economics*, 7, 185-208.

Bauer, P.W.(1990), "Rewcent Developments in the Econometric Estimation of Frontiers", Journal of Econometrics, 46, 39-56.

Bravo-Ureta, B.E. and A.E.Pinheiro (1993), "Efficiency Analysis of Developing Country Agriculture: A Review of the Frontier Function Literature", *Agricultural and Resource Economic Review*, 22, 88-101.

Bravo-Ureta, B.E. and A.E.Pinheiro (1993), "Efficiency Analysis of Developing Country Agriculture: A Review of the Frontier Function Literature", *Agricultural and Resource Economic Review*, 22, 88-101.

Coelli, T.J(1995b), "A Monte Carlo Analysis of the Stochastic Frontier Production Function", Journal of Productivity Analysis, 6, 247-268.

Coelli, T.J, Rao, S.P.D, O'Donnell. J, Battese, G.E (2005) "An Introduction to Efficiency and Productivity Analysis" *Springer Science*, New York, USA ISBN – 13: 978-0387-24266-8.

Coelli, T.J. (1995), "Estimators and Hypothesis Tests for a Stochastic: A Monte Carlo Analysis", *Journal of Productivity Analysis*, 6, 247-268.

Coelli, T.J. (1995a), "Recent Developments in Frontier Estimation and Efficiency Measurement", *Australian Journal of Agriculture Economics*, 39, 219-245.

Coelli, T.J.(1994) A Guide to FRONTIER *Version 4.1: A Computer Programme for Stochastic Frontier Production and Cost Function Estimation,* mimeo, Department of Econometrics, University of New England, Armidale, pp.32.

Coelli, T.J., (1996) A Guide to DEAP 2.1: "A Data Envelopment Analysis Computer Program", *CEPA working paper No.8/96,* ISBN 1863894969, Department of Econometrics, University of New England, Pp: 1-49

Farrell, M.J. (1957), "The Measurement of Productive Efficiency", *Journal of the Royal Statistical Society,* A CXX, Part 3, 253-290.

Greene, W. (2000), "Econometric Analysis", 4th Edition, *Prentice Hall,* Upper Saddle River, New Jersey.

Greene, W.H. (1993), "The Econometric Approach to Efficiency Analysis", in Fried, H.O., C.A.K. Lovell and S.S. Schmidt(eds), *The Measurement of Productive Efficiency: Techniques and Applications,* Oxford University Press, New York, pp. 68-119.

Huang, C.J. and J-T. Liu (1994), "Estimation of a Non-neutral Stochastic Frontier Production Function", *Journal of Productivity Analysis,* 4, 171-180.

Kalirajan, K.P and R.T. Shand(1994), "Economics in Disequilibrium: An Approach from Frontier", *Macmillan India Limited,* New Delhi.

Kalirajan, K.P. (1981), "An Econometric Analysis of Yield Variability in Paddy Production", *Canadian Journal of Agricultural Economics*, 29,283-294.

Kumbhakar, S.C., S.Ghosh and J.T.McGuckin (1991), "A Generalised Production Frontier Approach for Estimating Determinants of Inefficiency in U.S.Dairy Farms", *Journal of Business and Economic Statistics*, 9,279-286.

Kurma Charyulu D and Subho Biswas (2010), "Economics and Efficiency of Organic Farming vis-à-vis Conventional Farming in India" *Working Paper No. 2010-04-03,* CMA, IIM Ahmadabad, April 2010

Kurma Charyulu, D and Subho Biswas (2010), "Efficiency of Organic Input Units under NPOF Scheme in India" *Working Paper No. 2010-04-01,* CMA, IIM Ahmadabad, April 2010.

Lovell, C.A.K. (1993), "Production Frontiers and Productive Efficiency", in Fried, H.O., C.A.K. Lovell and S.S. Schmidt (Eds), *The Measurement of Productive Efficiency*, Oxford University Press, New York, 3-67.

Meeusen, W. and J. van den Broeck (1977), "Efficiency Estimation from Cobb-Douglas Production Functions with Composed Error", *International Economic Review*, 18, 435-444.

Parikh, A. and K.Shah(1994), " Measurement of Technical efficiency in the North West Frontier Provience of Pakistan", *Journal of Agricultural Economics,* 45, 132-138.

Ramsamy, C. *et al.,* (2003) "Hybrid Rice in Tamil Nadu Evaluation of Farmers' Experience" *Economic and political Weekly,* June 21 2003, pp.2509-2512.

Reifschneider, D. and R. Stevenson (1991), "Systematic Departures from the Frontier: A Framework for the Analysis of Firm Inefficiency", *International Economic Review,* 32, 715-723.

Sekar, C., C. Ramasamy and S.Senthilnathan (1994), "Size Productivity Relations in Paddy Farms of Tamil Naidu", *Agricultural Situation in India* 48: 859-863

Chapter 7

Summary, Conclusions and Policy Implications

It is a known fact that Agriculture is the backbone of the Indian Economy. Agriculture in India has a long history, dating back to 10,000 years. Today, India ranks second worldwide in farm output. Agriculture and allied sectors like forestry and logging accounted for 16 per cent of the GDP in 2010, employed 52 per cent of the total workforce and despite a steady decline of its share in the GDP, it is still the largest economic sector and plays a significant role in the overall socio-economic development of India[1]. India faced a severe food shortage when it was unshackled from the clutches of British rule and became independent in 1947. As a result, the Government gave primary importance to Agricultural Sector in the First Five Year Plan. Even then the situation continued till the 1960's. Then the Green Revolution has ushered in, in the Country, as a result of efforts of policy makers and agricultural scientists during mid 1960. This Programme aimed at attaining self-sufficiency in terms of food grains, empowering the farmers and modernizing agriculture by using modern techniques and tools to maximize the output of food.

The Green Revolution is one of the greatest triumphs of India. Within a decade, India completely stopped food imports from abroad and no longer was dependent on food aid from abroad. Even if there were food shortages in some parts of the Country, it never resulted in a famine. Thanks to the Green Revolution, India has now emerged as a notable exporter not only of food-grains, but also of several agricultural commodities. In spite of the advantages accrued to India, in terms of achieving self sufficiency in food production and increasing livelihood choices to the rural poor, Green Revolution made the Indian farmers and those world over to depend mostly on chemical fertilizers and pesticides, which degraded soil fertility, and environment.

The negative consequences of higher use of chemical fertilisers and pesticides are reduction in crop productivity and deterioration in the quality of natural resources. Pretty and Ball (2001)[2] have pointed out that the environment will be effected by the carbon emission of the agricultural system through: a) Direct use of fossil fuel in farm operations, b) Indirect use of embodied energy for producing agricultural inputs and c) Loss of soil organic matter during cultivation of soils.

Cole *et al.* (1997)[3] have observed that agriculture releases about 10-12 per cent of the total green house gasses emissions which is accounted for about 5.1 to 6.1 Gt CO_2. Joshi (2010)[4] has also pointed out that intensive agriculture and excessive use of external inputs are leading to degradation of soil, water and genetic resources and negatively effecting agricultural production. Arrouays and Pelissier(1994)[5]; Reicosky *et al.(*1995)[6],Sala and Paruelo(1997)[7]; Rasmussen *et al.*(1998)[8]; Tilman (1998)[9]; Smith(1999)[10] and Robert *et al.*(2001)[11], basing on the long term agrarian studies and experiments conducted in EU and North America have concluded that significant quantity of organic matter and soil corbon has been lost due to intensive cultivation

As a result of these changes in the agricultural sector, intellectuals world-over started searching for the ways to come out of the problem of heavy usage of chemical fertilizers and pesticides and finally arrived at to know that organic farming is the only remedy of the problem and also for sustainability of the Agricultural Sector in the long run. In this regard, Kramer *et al.*(2006)[12] pointed out that agriculture has the potential to reduce the emission of green house gasses by crop management agronomic practices. They pointed out that Nitrogen application rates in organic farming are 62-70 per cent lower than conventional agriculture due to recycling of organic crop reduce and use of manure. Some researchers have reported that yields of crops grown under organic farming system are comparable to those under conventional system. Nemecek *et al.* (2005)[13] have also reported that green house gasses emissions from organic farming are 36 per cent lower than conventional system of crop production. In addition, Regonald *et al*(1987)[14] and Siegrist *et al*(1998)[15] have reported that the organic farming system has the potential to improve soil fertility by retaining crop residues and reducing soil erosion. Niggli *et al.*(2009)[16] have reported that the organic farming system has the potential of reducing irrigation water and sequencing

CO_2. Mader *et al.* (2002)[17] and Pimental *et al.*(2005)[18] have observed that efficient use of inputs and net income per unit of cropped area on organic farms are at par due to reduction in costs of fertiliser and other input application. Reicosky *et al.* (1995)[19] and Fliessbach and Mader (2000)[20] have pointed out that the organic matter has a stabilizing effect on the soil structure, improves moisture retention capacity and protects soil against erosion. In this context, Pretty and Ball(2001)[21]; Niggly *et al*(2009)[22]have observed that organic farming has the potential to increase the sequestration rate on arable land and in combination with no tillage system of crop production, this can be easily increased by three to six quintal carbon per hectare per year.

As already noted, organic products are grown under a system of agriculture without any use of chemical fertilizers and pesticides with an environmentally and socially responsible approach. This is a method of farming that works at grass-roots level, preserving the reproductive and regenerative capacity of the soil, good plant nutrition, and sound soil management, produces nutritious food, rich in vitality and disease resistant.

The Problem

As already mentioned, of late, organic farming is gaining momentum in several advanced countries. India is no exception in this regard. Various studies on organic farming indicated that area and products covered under organic farming are increasing at a faster rate in advanced countries while its spread is relatively slow in developing countries like India. It is also evident that the growing demand for organic agricultural commodities in the advanced countries paves way for developing economies for potential export market for organic agricultural products. By international standards, conversion of a conventional farm into an organic farm will take a minimum of three years and during the first two years, the farmer may incur a loss in farming. In this context, a study of economics of organic farming in contrast to the conventional farming may throw light on the problems in the spread of organic farming. It is a fact that India is a developing country and most of the farmers are marginal and small holdings and are operating agriculture at subsistence levels. In this situation, a marginal or small farmer may not prefer to switch over to organic farming from his age old conventional farming due to the reasons mentioned above. But, if he is convinced of the economic benefits of organic farming, he readily accepts to switch over to organic farming. This fact was evident in the case of adoption of HYV seeds in the late 1960's. In turn, such types of studies may also help the policy makers to take appropriate measures to protect the farmer from economic losses in this process of conversion.

Need for the Study

Of late, many advanced countries like the USA, Switzerland, Australia, Western Europe etc evinced interest in the organic farming practices which generally assure sustainability of agriculture also to the next generation without any compromise on the food needs of the present generation in particular and natural resources like land, water, and environment in general. It is argued that for sustainability of agricultural sector of any country, organic farming is the only way-out as it assures no contamination of water, no environmental pollution and no degradation of soil fertility.

With this back-ground, it can be concluded that there is an urgent need to address this problem in a holistic approach to encourage farmers at the grassroots level to take up organic farming. Also a review of literature has revealed that except the pioneering works on organic farming at the CMA[23], IIM, Ahmadabad, which confined their attention to the Northern and Western parts of India, on paddy, wheat, sugarcane and cotton and on the efficiency of inputs used in organic farming and conventional farming and another peripheral study by Prasad[24] which studied several comparative aspects of organic farming and conventional farming, no researcher in India has so far examined location-specific and crop-specific aspects relating to economics of organic farming in a State.

Hence, a comprehensive study dealing with the economics of organic farming and conventional farming covering different agro-climatic conditions is felt necessary. As such, the present Study addressed itself to fill in this gap by examining the Economics of Organic Farming vis-à-vis Conventional Farming in A.P. covering paddy, redgram and groundnut among cereals, pulses and oil-seeds in East Godavari, Mahabubnagar and Anantapur respectively. An attempt has been made in this Study to examine the Economics of Organic Farming in Andhra Pradesh with the following objectives:

Objectives

The main objectives of this Study are:

1. To examine the trends in the area, production and productivity of the selected crops viz. paddy, redgram and groundnut in the State of Andhra Pradesh and the selected districts of Andhra Pradesh,
2. To analyse the cost of and returns from organic farming practices vis-à-vis conventional farming practices,
3. To assess the economic efficiency of organic farming over conventional farming through the estimation of technical efficiency and allocative efficiency,

4. To identify the factors determining technical efficiency and
5. To suggest measures that may be useful to the policy makers both at the micro and macro levels.

Methodology an d Sample Design

This Study is based on both primary and secondary data collected from various sources. The sample households for collection of primary data have been selected by using the multi stage stratified random sampling technique. The State of Andhra Pradesh is the study area and three major crops, one each from cereals, pulses and oilseeds viz., paddy, redgram and groundnut have been selected basing on the proportion of area under organic farming. Among the 23 districts of Andhra Pradesh, East Godavari, Mahabubnagar and Anantapur have been selected as they are predominantly cultivating the selected crops under organic farming respectively, which also represent the three natural/geographical regions of Andhra Pradesh viz., Coastal Andhra, Telangana and Rayalaseema. In the second stage 250 paddy cultivating households comprising of 150 organic farmers and 100 conventional farmers' households have been selected from East Godavari District. From Mahabubnagar District, 150 Redgram cultivating households comprising 100 from organic farmers and 50 from conventional farmers households have been selected From Anantapur District 150 Groundnut cultivating households comprising 100 from organic farmers and 50 from conventional farmer households have been selected. The selection of sampling units in each district for each crop is based on the stratified random sampling technique. A pre-tested and well designed schedule has been canvassed among the selected sample holdings to elicit information on structure of farm holdings, demographic characteristics, asset structure, cost of cultivation, returns etc. The secondary data have been collected from various issues of Statistical Abstract of Andhra Pradesh and Season and Crop Reports being published annually by the Directorate of Economics and Statistics, Govt. of Andhra Pradesh. The reference year of the Study is 2010-11.

TECHNIQUES USED

Simple statistical tools like averages and percentages have been used in analysing the collected data. Further, Stochastic Frontier Production Function (SFPF) 4.1 and Data Envelopment Analysis (Computer) Programme (DEAP) 2.1 techniques have been employed to assess technical efficiency and allocative efficiency under various situations. In addition, multiple regression analysis has been used to identify the factors determining technical efficiency.

THE STOCHASTIC FRONTIER AND EFFICIENCY MODEL

The Stochastic Frontier Production Function (SFPF) Model which specify for the farming operations in a given farm category is as follows:

$$\ln(Y) = \beta_0 + \beta_1 \ln(Land) + \beta_2 \ln(HL) + \beta_3 \ln(TL) + \beta_4 (Seed) + \beta_5 \ln(OF) + \beta_6 \ln(OP) + \beta_7 \ln(CS) + U - V$$

ln represents the natural logarithm(i.e., to base e).

Y represents the total value of output (in '₹') from the crop which are grown.'

Land represents the total area of cropped land (in Acres)

HL represents the total quantity of human labour (family and hired labourers) measured in value terms (in '₹').

TL represents the total amount of bullock and machine labour (owned and hired) (in '₹').

Seed represents value of Seed (in '₹').

OF represents amount of organic fertilisers (in '₹') in case of organic farms and amount of chemical fertilizers in case of conventional farms.

OP represents the amount of organic pesticides (in '₹') in case of organic farms and amount on synthetic pesticides in case of conventional farms.

CS represents the amount on capital services (in '₹') which include depreciation on farm assets and interest on working capital.

The *Vs* are assumed to be independently and identically distributed random errors, having $N\ (0,\ \acute{o}^2)$ distribution; and

The *Us* are non-negative random variables, called technical efficiency effects, associated with the technical efficiency of production of the farmers involved.

The technical efficiencies under the above mentioned assumptions for organic farmers and conventional farmers and the parameters of the Model are estimated by the method of maximum likelihood, using the computer program, FRONTIER version 4.1 Coelli, (1992, 1994)[25]

THE DEA MODEL:

The gross value of production per acre as output (Y) and costs on five inputs, viz., human labour (X_1), traction labour (X_2), seed (X_3), fertiliser (X_4) and pesticides (X_5). Input oriented DEA Model is used and the analysis is carried out by using Data Envelopment Analysis (Computer) Programme (DEAP) 2.1 Coelli, (1996)[26].

FACTORS DETERMINING TECHNICAL EFFICIENCY MODEL:

Multiple regression model for the factors determining technical efficiency of farmers:

$TE_i = \beta_0 + \beta_1 X_1 + \beta_2 X_2 + \beta_3 X_3 + \beta_4 X_4 + \beta_5 D_1 + \beta_6 D_2 + e_t$

TE_i = Technical Efficiency of i-th farmer

X_1 = Age of the farmer in years

X_2 = Years of schooling of the farmer

X_3 = Distance to the market (in Kilometers)

X_4 = Experience of the farmer in farming years (for organic farmers)

D_1 = 1, if belongs to small farmers
0, if otherwise

D_2 = 1, if belongs to small farmers
0, if otherwise

e = error term

β_1 = regression coefficients to be estimated (i = 0, 1........6)

LIMITATIONS OF THE STUDY

Due to paucity of time and resources, survey method has been adopted to collect relevant information, using schedules designed for the purpose by personal interview. The necessary data were obtained basing on the recall/memory of the farmers which has many inherent limitations. Peasants do not maintain accounts and do not generally disclose them even if they do. But care has been taken to crosscheck the accuracy of the data. Since the results were based on the data pertaining to only one agricultural year i.e., 2010 – 11, the application of the results should be done with due care.

In addition, the nature of data used in this Study has certain limitations. The data relate to one year and pertain to an agriculturally developed district of Andhra Pradesh, East Godavari, which is a rice granary of Andhra Pradesh, while Mahabubnagar and Anantapur are predominately redgram and groundnut growing areas respectively. Time series data, giving a comparative picture of the same farm over a period of time would better serve the objectives of the Study. The price data relating to crop output is represented by farm harvest prices. Similarly, the prevailing market prices of different farm inputs at the time of investigation are considered. This is mainly due to the adoption of survey method of data collection. However, cost accounting method may give better and meaningful insights. This is also another limitation of the Study.

ISSUES FOR FURTHER RESEARCH

Several issues for further research have been identified through a review of literature and several among many such issues have been listed below:

- Impact Assessment of organic farming in different Eco-regions.
- Sustainability of organic farming with respect to Environment up-gradation in a specific region.
- Adoption of organic farming and location-specific constraints.
- Institutional and policy issues of organic farming in a different presentation.

MAJOR FINDINGS

- The literacy levels of East Godavari District are higher for both males and females compared to Anantapur and Mahabubnagar.
- While more than one half of the population of Andhra Pradesh, East Godavari and Anantapur Districts is unproductive, it is lower in Mahabubnagar.
- The major source of irrigation in East Godavari District is canals, which constitutes 49 per cent of the total operated area of the District, while in Mahabubnagar and Anantapur district, tube well / dug well, constitute 18 per cent and 8 per cent of the total operated area respectively. The State figures indicate that tube wells / dug wells irrigate about 16 per cent to total operated area, followed by canals (12 per cent).
- The percentage of buffaloes in the total live-stock population is very high in East Godavari District, while in Anantapur and Mahabubnagar districts; the percentage of sheep to the total livestock population is high constituting 83 per cent and 58 per cent respectively.
- The socio-economic profile of the study area reveals that the conditions prevailed in East Godavari District like literacy rate, percentage of the aged and experienced population in to total population, average rain-fall, irrigation facilities and availability of dung (organic manure), are more favorable for organic farming compared to the other selected districts. Thus, it can be concluded that East Godavari District is congenial for organic farming compared to the other two selected districts. So, it can be hypothesized that the organic farmers in East Godavari District are in an advantageous position in relation to efficient input-use compared to other farmers in Mahabubnagar and Anantapur.

- An analysis on demographic profile/characteristics has revealed that there is not much of difference in both the categories of farms viz., organic and conventional, like age, gender, family size etc., and economic characteristics like value of assets', size of land holding etc. Both the categories of farms can be differentiated with regard to the various levels of literacy, as the percentage of farmers with secondary and higher levels of education is more in organic farming category compared to their counterparts. As a result, it can be hypothesized that the farmers of organic farming category are more rational, have more accessibility to the information on organic farming practices, which consequently leads to efficient input-use.
- The cost of paddy per farm and per acre on the basis of different cost concepts is found to be relatively higher on conventional farms compared to organic farms. The same phenomenon is discernible among different size groups of farms also.
- Further, it is to be noted that in case of organic paddy holdings, the proportions of different costs to Cost – C_2 and farms size are directly related, whereas in case of conventional holdings, the proportions of different costs to Cost – C_2 are inversely related.
- The cost of redgram per farm and per acre on the basis of different cost concepts is found to be relatively higher on conventional farms compared to organic farms. The same phenomenon is discernible among different size groups of farms also. Further, it is to be noted that in case of organic holdings, the proportions of different costs to Cost – C_2 are directly related, whereas in case of conventional holdings, the proportions of different costs to Cost – C_2 are inversely related.
- The cost of groundnut per acre on the basis of different cost concepts is found to be relatively higher on conventional farms compared to organic farms. Except the small farm holdings, the same phenomenon is discernible among different size groups of farms also and the cost of cultivation for small farm holdings on organic farming is slightly higher. In case of organic holdings, the proportions of different costs to Cost – C_2 and farms size are inversely related, whereas in case of conventional holdings, these proportions are directly related.
- The farmers of all size groups of the selected crops under both organic and conventional category are getting profits, but the profits earned by the organic farmers are higher by 37 per cent, 33 per cent and 59 per cent for the selected crops respectively. A more or less similar picture can be seen from the analysis of different size groups of farms on both the organic and conventional category of the selected crops except for the small farmers of redgram. The small farmers of organic

redgram are getting lower profits or net incomes than their counterpart by ₹ 47/- (0.77 per cent), which is a very negligible amount.

- When perceptions of organic farmers were elicited as to their experiences in organic farming, certification, problems they encounter with etc., it is heartening to note that as many as 18 per cent have been adopting organic farming since 2001 and all of them have been continuing it to date. Despite this fact, about 15 per cent of them have switched over to organic farming only in 2005 and all the selected organic farmers have crossed the gestation period of three years and have been reaping the benefits of organic farming.
- Electronic media has more impact on the switching over to organic farming, as it is evident from the fact that it motivated around 21 per cent of the total sample farmers, followed by village cooperative (19 per cent), print media (17 per cent), village leaders (15 per cent), agricultural extension workers (15 per cent) and fellow farmers (13 per cent). Slight variations in the percentages, can be found at the crop level analysis also.
- It is distressing to note that out of the selected organic farmers none has reported that he has obtained certification, though many of them have reported that they have taken organic farming as early as in 2001. The sample farmers of the study area based on their experience in organic farming reported some advantages of organic farming which are correlated with the results of the earlier studies. Around 34 per cent of them reported that the fertility of soil has increased. In addition, around 37 per cent of them reported that the cost of cultivation has come down due to non-usage of chemical fertilisers. Further, around 15 per cent of them reported that the organic produce is good for health, while another 13 per cent of them reported that they are getting higher and constant returns from organic farming.
- With regard to the certification for organic produce, they expressed, that certification is highly expensive (66 per cent), followed by lack of information on the certification process (27 per cent) and small size of farm holdings (7 per cent).
- When information was elicited as to other problems, almost all of them reported that they have been facing problems in marketing their produce as their product lacks certification.
- All the sample farmers suggested that the organic farming will spread if the government provides some subsidies on organic inputs and support them in getting certification and enable them to market their produce at remunerative prices. In addition, they suggested that any technical support from the agricultural line department will also be of

quite help for them. As a whole, the farmers felt that it is in the hands of government to encourage the organic farming on a wider scale.

- Both organic and conventional farmers are getting benefited with regard to the various standard concepts of returns employed and analysed in this Study. It can also be seen that the small farmers of organic redgram, are getting lower profits compared their counterparts. Another important observation that can be made from the analysis is that organic groundnut farmers of large farm size group are getting lower profits compared their counterparts. Based on these conclusions, it could not be generalised that the organic farmers are more efficient both technically and allocatively compared to the conventional farmers.
- Results of the Stochastic Frontier Production Function Model indicated that technical efficiency is relatively higher on organic paddy farms compared to conventional paddy farms, while conventional redgram farmers are more efficient compared to their counterparts and organic groundnut farmers are relatively more efficient than their counterparts.
- Analysis of CRS-TE, VRS-TE and SE Model concluded that both organic paddy and groundnut farms operate close to their specific frontiers than conventional farms, while conventional redgram farms operate close to their specific frontier than organic redgram farms.
- Analysis of TE, AE and EE - CRS Model concluded that organic paddy and groundnut farms are more efficient compared to conventional paddy and groundnut farms, while conventional redgram farms are more efficient compared to organic redgram farms.
- Similarly analysis of TE, AE and EE - VRS Model concluded that organic paddy and groundnut farms are more efficient compared to conventional paddy and groundnut farms, while conventional redgram farms are more efficient as compared to organic redgram farms.
- Further an analysis of Factors Determining Technical Efficiency Model concluded that age of the farmer, education, distance to market and experience of the farmer appear to be predominant variables determining technical efficiency and large farms are found to be more efficient than medium and small farms.

POLICY IMPLICATIONS

- Most studies have found that organic agriculture requires significantly greater labour input compared to conventional farms. Therefore, the diversification of crops typically found on organic farms, with their different planting and harvesting schedules, may distribute labour demand more evenly, which could help stabilize employment. As in all agricultural systems, diversity in production increases income-generating opportunities and can, as in the case of fruits, which supply the essential health-protecting minerals and vitamins for the family diet. It also spreads the risks of failure over a wide range of crops.

- Several studies have argued that for sustainability of agricultural sector of any country, organic farming is the only way-out as it assures no contamination of water, no environmental pollution and no degradation of soil fertility.
- A study in Egypt has concluded that the quality of drinking water will improve further with an expected expansion of organic agriculture and organic agriculture enables ecosystems to better adjust to the effects of climate change and has a major potential for reducing agricultural greenhouse and other gas emissions.
- It is well known that organic and integrated systems had higher soil quality and potentially lower negative environmental impact than the conventional system. When compared with the conventional and integrated systems, the organic system produced sweeter fruit, higher profitability, greater energy efficiency and further indicated that the organic system ranked first in environmental and economic sustainability, while the integrated system ranked second and the conventional system last.
- As per a study, India needs at least 294 million tonnes of food-grain per annum by 2020 and the mainstream of Indian agriculture has to depend on modern agricultural inputs, such as chemical fertilizers and pesticides. Nevertheless, their restrained and efficient use is important. As regards plant nutrient needs in modern agriculture, the Study suggested that integrated nutrient supply is the key for the sustainability of Indian agriculture.

In this context, the role of the government is critical in motivating farmers towards organic farming in the Country. Some of the major suggestions for expansion of organic farming are:

- Creation of separate 'green channels' for marketing of organic foods.
- Announcement of premium prices for organic staple food crops in advance of crop season.
- Creation of demand by more awareness programmes.
- Provision of Input/conversion subsidies for encouraging organic growers.

- Investment of more funds on Research and Development on organic farming, Initiation of cheaper and quicker certification process for organic produce.
- Farmers in the Study area reported that they are not getting any assistance whatsoever either from the Agricultural Department or from other government agencies. As such, the intervention of NGO's is very much needed in this regard.

Notes

1 Economic Survey 2011, *Planning Commission*, Government of India and for a detailed discussion on the general economic development of India in the recent past, see for instance, Mohana Rao. L.K, budget Meet 2011 held at Dept. of Economics, Andhra University on 5th April 2011.

2 Pretty, Jules and Ball Andrew (2001), Agricultural Influences on Carbon Emissions and Sequestration: A Review of Evidence and the emerging Trading Options, *Occasional Paper, Centre for Environment and Society and Department of Biological Sciences*, University of Essex, U.K.

3 Cole, C.V.; J. Duxbury, J. Freney, O. Heinemeyer, K. Minami, A. Mosier, K. Paustin, N. Rosenberg; N. Sampson, D. Sauerbeck and Q. Zaho (1997), "Global Estimates of Potential Mitigation of Greenhouse Gas Emissions by Griculture," *Nut Cycl Agroecosyst*, Vol. 49, pp. 221-228.

4 Joshi. P.K., (2010) "Conservation Agriculture: An Overview", *Indian Journal of Agricultural Economics*, Vol.66, No.1 pp.53-63.

5 Arrouays, D. and P.Pelissier (1994), "Changes in Carbon Storage in Temperate Humic Soils After Forest Clearing and Continuous Corn Cropping in France", *Plant Soil*, Vol.160, pp.215-223.

6 Reicosky, D.C, W.D. Kemper, G. W. Langdale, C.L. Douglas and P.E. Rasmussen (1995), "Soil Organic Matter Changes Resulting From Tillage and Biomass Production," *Journal of Soil and Water Conservation*, Vol.50, No.3, pp.253-261.

7 Sala, O.E. and J.M. Paruelo (1997), "Ecosystem Services in Grasslands", in G. Daily (Ed) (1997), *Nature's Services: Societal Dependence on Natural Ecosystems*, Island Press, Washington, D.C., U.S.A.

8 Rasmussen, P.E., K.W.T. Goulding, J. R. Brown, P. R. Grace, H.H. Janzen and M. Korschens (1998), "Long Term Agro-ecosystem Experiments: Assessing Agricultural Sustainability and Global Change", *Science*, Vol.282, pp.893-896.

9 Tilman, D. (1998), "The Greening of the Green Revolution", *Nature,* Vol.396, pp.211-212.

10 Smith, K.A. (1999), "After Kyoto Protocol: Can Scientists Make a Useful Contribution?" *Soil Biol. Biochemistry,* Vol.15,pp.71-75.

11 RobertM., J. Antoine and F. Nachtergaele (2001), *Carbon Sequestration in soils, Proposal for Land Management in Arid Areas of the Tropics,* AGLL, Food and Agriculture Organization of the United Nations, Rome, Italy.

12 Kramer, S.B.; J.P. Reganold; J.D. Glover; B.J.M. Bohannan H. A. mooney (2006), " Reduced Nitrate Leaching and Enhanced Denitrifier Activity and Efficiency in Organically Fertilised Soils" *Proceedings of the National Academy of Sciences of the USA.*, Vol. 103, pp. 4522-4527

13 Nemecek, T; O. Hugnenin. Elie, D. Dubois and G. Gailord (2005) "Okobilanzierung von anbausystemen im schweizericschen Acker – und futterbau", *Schriftenreihe der* FAL, 58 FAL Reckenholz, Zurich

14 Regonald, j.P,; L.F. Elliot and Y.L. Unger (1987), Long-Term Effects of Organic and Conventional Farming on Soil Erosion", *Nature,* Vl.330, pp.370-372

15 Siegrist, S., D. Staub, L. Pfiffner and P. Mader (1998) "Does Organic Agriculture Reduce Soil Erodibility? The Results of a Long-Term Field Study on Losses in Switzerland," *Agriculture, Ecosystems and Environment,* Vol.69, pp. 253-264.

16 Niggli, U., A. Fliebach, P. Hepperly, J. hanson, D. Douds and R. Seidel (2009), "Low Greenhouse Gas Agriculture: Mitigation and Adoption Potential of Sustainable Farming System", *Food and Agriculture Organization, Review – 2,* pp.1-22.

17 Mader, P., A. Fliebach, D. Dubois, L. Gunst, P. Fried and U. Niggli (2002), "Soil Fertility and Biodiversity in Organic Farming", *Science,* Vol.296,pp.1694-1697.

18 Pimentel, D., P. Hepperly, J. Hanson, D. Douds and R. Seidel (2005), "Environmental, Energetic and Economic Comparisons of Organic and Conventional Farming Systems", *Bioscience,* Vol.55 pp.573-582.

19 *Op. cit*

20 Fliessbach, A. and P. Mader (2000), "Microbial Biomass and Size-Density Fractions Differ Between Soils or Organic and Conventional Agriculture Systems", *Soil Biol. Biochemistry,* Vol.32,pp. 757-768.

21 *Op. cit.*

22 *Op. cit.*

23 Kurma Charyulu D and Subho Biswas (2010), "Economics and Efficiency of Organic Farming vis-à-vis Conventional Farming in India" *Working Paper No. 2010-04-03,* CMA, IIM Ahmadabad, April 2010

24 Prasad, R. (1999), Organic farming *vis-à-vis* modern agriculture *Curr. Sci.,* 1999, 77, 38–43.

25 Coelli, T.J.(1994) A Guide to FRONTIER *Version 4.1: A Computer Programme for Stochastic Frontier Production and Cost Function Estimation,* mimeo, Department of Econometrics, University of New England, Armidale, pp.32.

26 *Op.cit.*

Select Bibilography

Abouleish Helmy (2007) "Organic agriculture and food Utilisation - an Egyptian case study" Managing Director, SEKEM Group, Egypt, available on the world wide web :ftp://ftp.fao. org/paia /organicag/ofs/07-Abouleish.ppt

Acs S., P.B.M. Berentsen, R.B.M. Huirne (2006) "Conversion to organic arable farming in The Netherlands: A dynamic linear programming analysis" Published in *"ELSEVIER"* and also available on the world wide web: http://doi:10.1016/j.agsy.2006.11.002

Aigner, D.J., Lovell, C.A.K. and Schmidt, P. (1977), Formulation and Estimation of Stochastic Frontier Production Function Models", *Journal of Econometrics*, 6, 21-37.

Ali, A.I. and Seiford, L. (1993) "The Mathematical Programming Approach to Efficiency Analysis", in (Eds.) H.O Fried, C.A.K Lovell and S.S Schmidt, *The Measurement of Productive Efficiency*, Pp: 129-159, *Oxford University Press*, New York

Anand Raj Daniel, K. Sridhar, Arun Ambatipudi, H. Lanting, & S. Brenchandran,(2005)Second "International Symposium on Biological Control of Arthropods", available on the world wide web: http://www.bugwood.org /arthropod2005/ vol1/6c.pdf

Anderson M.D (1994) "Economics of organic farming in USA" in *The economics of organic farming – An international perspective (ed.)* by Lampkin N.H and Padel S., CAB International Publishers

Anderson, J.C., Wachenheim, C.J., & Lesch, W.C. (2006). "Perceptions of genetically modified and organic foods and processes". *AgBioForum*, 9(3), 180-194. available on the world wide web: http://www.agbioforum.org

Arrouays, D. and P.Pelissier (1994), "Changes in Carbon Storage in Temperate Humic Soils After Forest Clearing and Continuous Corn Cropping in France", *Plant Soil,* Vol.160, pp.215-223.

Auigner,D.J. and S.F. Chu (1968), " On Estimating the Industry Production Function", *American Economic Review* 58, 826-839.

Bagi, S.F. (1981), "Relationship between Farm Size and Economic Efficiency: An Analysis of Farm Level Data from Haryana (India)", *Canadian Journal of Agricultural Economics* 29:317-326.

Balk, B.M (2001) "Scale Efficiency and Productivity Change", *Journal of Productivity Analysis",* 15, 159-183

Battese G.E, Coelli T.J (1988) Prediction of Grm-level technical efficiencies: With a generalized frontier production function and panel data, *Journal of Econometrics* 38:387-399

Battese G.E, Coelli T.J (1992) "Frontier production functions, technical efficiency and panel data with application to paddy fanners in India", *Journal of Productivity Analysis* 3:163-169

Battese G.E, Coelli T.J (1993) A stochastic frontier production function incorporating a model for technical inefficiency effects, *Working Papers in Econometrics and Applied Statistics No 69,* Department of Econometrics, University of New England. Armidale

Battese, G. E. and Rao P. D. S. (2002) "Technology Gap, Efficiency, and a Stochastic Metafrontier Function" *International Journal of Business and Economics,* 2002, Vol. 1, No. 2, 87-93

Battese, G. E. and Coelli, T.J (1995) "A Model for Technical Inefficiency Effects in a Stochastic Frontier Production Function for Panel Data" *Empirical Economics* (1995) 20, pp. 325-332

Bauer PW (1990) Recent developments in the econometric estimation of frontiers, *Journal of Econometrics* 46:39-56

Bauer, P.W.(1990), "Rewcent Developments in the Econometric Estimation of Frontiers", Journal of Econometrics, 46, 39-56.

Bravo-Ureta, B.E. and A.E.Pinheiro (1993), "Efficiency Analysis of Developing Country Agriculture: A Review of the Frontier Function Literature", *Agricultural and Resource Economic Review, 22,* 88-101.

Cembalo, L., and Cicia, G (2002) "Disponibilita di dati ed opporunita di analisi del database RICA biologico. In Scardera, A., Zanoli, R. (ed)" *L'agricoltura biologica in Italia. Rome*: INEA, 63-76.

Charnes, A., W.W. Cooper, A.Y Lewin and L.M Seiford (1995) "Data Envelopment Analysis: Theory, Methodology and Applications, *Kluwer Publications.*

Coelli T.J and S.Perelman (1996) A comparison of parametric and Non-parametric Distance functions: with applications to European Railways, CREPP *Discussion Paper,* University of Liege, Liege.

Coelli TJ (1992) A computer program for frontier production function estimation: FRONTIER Version ZO, *Economics Letters* 39:29-32

Coelli, T.J and G. E. Battese (1996) "Identification of Factors which Influence the Technical Inefficiency of Indian Farmers" *Australian Journal of Agricultural Economics,* Vol. 40, No.2 (August 1996), pp.103-128.

Coelli, T.J(1995b), "A Monte Carlo Analysis of the Stochastic Frontier Production Function", Journal of Productivity Analysis, 6, 247-268.

Coelli, T.J, Rao, S.P.D, O'Donnell. J, Battese, G.E (2005) "An Introduction to Efficiency and Productivity Analysis" *Springer Science,* New York, USA ISBN – 13: 978-0387-24266-8.

Coelli, T.J. (1995), "Estimators and Hypothesis Tests for a Stochastic: A Monte Carlo Analysis", *Journal of Productivity Analysis,* 6, 247-268.

Coelli, T.J. (1995a), "Recent Developments in Frontier Estimation and Efficiency Measurement", *Australian Journal of Agriculture Economics,* 39, 219-245.

Coelli, T.J.(1994) A Guide to FRONTIER *Version 4.1: A Computer Programme for Stochastic Frontier Production and Cost Function Estimation,* mimeo, Department of Econometrics, University of New England, Armidale, pp.32.

Coelli, T.J., (1996) A Guide to DEAP 2.1: "A Data Envelopment Analysis Computer Program", *CEPA working paper No.8/96,* ISBN 1863894969, Department of Econometrics, University of New England, Pp: 1-49

Coelli, TJ., (1995) "Recent Developments in Frontier Modelling and Efficiency Measurement, *Australian Journal of Agricultural Economics,* 39, Pp: 219-245

Coelli, TJ., Rao, P.D.S and Battese, G.E (2002) "An Introduction to Efficiency and Productivity Analysis", *Kluwer Publishers,* London

Cole, C.V.; J. Duxbury, J. Freney, O. Heinemeyer, K. Minami, A. Mosier, K. Paustin, N. Rosenberg; N. Sampson, D. Sauerbeck and Q. Zaho (1997), "Global Estimates of Potential Mitigation of Greenhouse Gas Emissions by Griculture," *Nut Cycl Agroecosyst,* Vol. 49, pp. 221-228.

Delate Kathleen and Cynthia A. (2003), "Organic Production Works", Online, *Institute of Science in Society,* available on the world wide web: http://lists.ifas.ufl.edu/cgi-bin/wa.exe?A2=ind0412&L=sanet13

Demand for organic products has created new export opportunities for the developing world, Spotlight / 1999, This article is based on Organic agriculture, a report to the *FAO Committee on Agriculture (COAG)*, which met in Rome on 25-26 January 1999.

Dimitri Carolyn and Catherine Greene(, Recent Growth Patterns in the U.S. Organic Foods Market U.S. Department of Agriculture, Economic Research Service, Market and Trade Economics Division and Resource Economics Division. Agriculture Information Bulletin Number 777. available on world wide web: http://www.ers.usda.gov/publications/aib777/aib777a.pdf

Dubgaard A (1994) "Economics of organic farming in Denmark" in The economics of organic farming – An international perspective (ed) by Lampkin N.H and Padel S., *CAB International Publishers*

Economic Survey (2010-11) *Planning Commission of India*, Government of India, New Delhi.

Fare, R and D. Primont (1995) "Multi-output production and Duality: Theory and Applications, *Kluwer Academic Publishers*, Boston.

Fare, R., S. Grosskopf and P.Roos (1998), "Malmquist Productivity Indexes: A survey of Theory and Practice", in R.Fare, S. Grosskopf and R.R Russell (Eds.) Index Numbers: Essays in Honour of Sten Malmquit, *Kluwer Academic Publishers*, Boston.

Farrell, M.J (1957) Generalized Farrell Measures of Efficiency: An application to Milk processing in Swedish Dairy plants. *"Economic Journal"*, Vol no.89, Pp: 294-315

Fersund FR, Lovell CAK, Schmidt P (1980) A survey of frontier production functions and their relationship to efficiency measurement, *Journal of Econometrics* 13:5-25

Fliessbach, A. and P. Mader (2000), "Microbial Biomass and Size-Density Fractions Differ Between Soils or Organic and Conventional Agriculture Systems", *Soil Biol. Biochemistry*, Vol.32,pp. 757-768.

Funtanilla M., Lyford C. and Wang C. (2009) "An evaluation of organic cotton marketing opportunity", Paper prepared and presented at the Agricultural & Applied Economics Association 2009, *AAEA & ACCI Joint Annual Meeting*, Milwaukee, WI, July 26-28, 2009.

Good, D.H., M.I Nadiri, L.H Roller and R.C Sickles (1993), "Efficiency and Productivity growth comparisons of European and US Air carriers: A First look at the data". *"Journal of Productivity Analysis"* vol no.4, June, Pp. 115-125

Greene, W. (2000), "Econometric Analysis", 4th Edition, *Prentice Hall*, Upper Saddle River, New Jersey.

Greene, W.H. (1993), "The Econometric Approach to Efficiency Analysis", in Fried, H.O., C.A.K. Lovell and S.S. Schmidt(eds), *The Measurement of Productive Efficiency: Techniques and Applications*, Oxford University Press, New York, pp. 68-119.

Gu¨ndog¢mus Erdemir (2006) "Energy use on organic farming: A comparative analysis on organic versus conventional apricot production on small holdings in Turkey" Published in *"ELSEVIER"*. and also available on the world wide web http:// doi:10.1016/j.enconman.2006.01.001

Heady. Earl.O and John L. Dillon,(1965) "Agricultural Production Functions" ISBN-87-7096-129-7, Kalyani Publishers, New Delhi.

Huang, C.J. and J-T. Liu (1994), "Estimation of a Non-neutral Stochastic Frontier Production Function", *Journal of Productivity Analysis*, 4, 171-180.

John Henning (1994) "Economics of organic farming in Canada" in The economics of organic farming - An international perspective (ed) by Lampkin N.H and Padel S., *CAB International Publishers*

Joshi. P.K., (2010) "Conservation Agriculture: An Overview", *Indian Journal of Agricultural Economics*, Vol.66, No.1 pp.53-63.

Kaiser, J. (2004) Wounding earth's fragile skin, Science, 304, p 1616-1618 Lampkin, N.H (1994) "Organic farming: sustainable agriculture in practice" in Lampkin N.H and Padel, S (eds.) The economics of organic farming – An international perspective, *CAB International*, Oxon (UK)

Kalirajan, K.P and R.T. Shand(1994), "Economics in Disequilibrium: An Approach from Frontier", *Macmillan India Limited*, New Delhi.

Kalirajan, K.P. (1981), "An Econometric Analysis of Yield Variability in Paddy Production", *Canadian Journal of Agricultural Economics*, 29,283-294.

Kassie, Menale, Zikhali, Precious, Pender, John and Köhlin, Gunnar(2008) "Organic Farming Technologies and Agricultural Productivity: The case of Semi-Arid Ethiopia" Paper provided by Göteobrg University, Dept. of Economics in its series Working Papers in Economics with number 334, 2008, available on the world wide web:http://ideas.repec.org/p/hhs/gunwpe/0334.html.

Kirchmann Holger, Lars Bergström, Thomas Kätterer, Lennart Mattsson and Sven Gesslein (2007), "Comparison of Long-Term Organic and Conventional Crop–Livestock Systems on a Previously Nutrient-Depleted Soil in Sweden", Published in *Agron* J 99:960-972 (2007), American Society of Agronomy, 677 S. Segoe Rd., Madison, WI 53711 USA, available on the world wide web: http://agron.scijournals. org/cgi/content/ abstract/ 99/4/960

Kramer, S.B.; J.P. Reganold; J.D. Glover; B.J.M. Bohannan H. A. mooney (2006), " Reduced Nitrate Leaching and Enhanced Denitrifier Activity and Efficiency in Organically Fertilised Soils" *Proceedings of the National Academy of Sciences of the USA.*, Vol. 103, pp. 4522-4527

Kumbhakar, S.C., S.Ghosh and J.T.McGuckin (1991), "A Generalised Production Frontier Approach for Estimating Determinants of Inefficiency in U.S.Dairy Farms", *Journal of Business and Economic Statistics*, 9,279-286.

Kurma Charyulu D and Subho Biswas (2010), "Economics and Efficiency of Organic Farming vis-à-vis Conventional Farming in India" *Working Paper No. 2010-04-03,* CMA, IIM Ahmadabad, April 2010

Kurma Charyulu, D and Subho Biswas (2010), "Efficiency of Organic Input Units under NPOF Scheme in India" *Working Paper No. 2010-04-01,* CMA, IIM Ahmadabad, April 2010.

Lampkin N H (1994) "Economics of organic farming in Britain" in The economics of organic farming – An international perspective (ed) by Lampkin N.H and Padel S., *CAB International Publishers*

Larsen K., Foster K., (2005) "Technical Efficiency among Organic and Conventional Farms in Sweden 2000-2002: A Counterfactual and Self Selection Analysis". Paper presented at the *American Agricultural Economics Association Annual Meeting,* Providence, Rhode Island, July 24-27.

Lesjak Heli Annika (2008) "Explaining organic farming through past policies: comparing support policies of the EU, Austria and Finland" *Journal of Cleaner Production 16 (2008) 1- 11* and also available on world wide web: http:// doi:10.1016/j.jclepro.2006.06.005

Lovell, C.A.K. (1993), "Production Frontiers and Productive Efficiency", in Fried, H.O., C.A.K. Lovell and S.S. Schmidt (Eds), *The Measurement of Productive Efficiency,* Oxford University Press, New York, 3-67.

Madau F., (2005) "Technical efficiency in organic farming: an application on Italian cereal farms using a parametric approach", *Proc. XI Congress EAAE,* August 24-27, Copenhagen, Denmark

Mader Paul, *et. al.* (2002), "Soil Fertility and Biodiversity in Organic Farming", *Science* 31 May 2002: Vol. 296. no. 5573, pp. 1694 – 1697, DOI: 10.1126/science.1071148, available on the world wide web: http://www.sciencemag.org/cgi/content/abstract/296/5573/1694

Mader, P., A. Fliebach, D. Dubois, L. Gunst, P. Fried and U. Niggli (2002), "Soil Fertility and Biodiversity in Organic Farming", *Science,* Vol.296,pp.1694-1697.

Mayen C.D, Balajgtas J.V and Alexander C.E (2010) "Technology Adoption and Technical Efficiency: Organic and Conventional Dairy Farms in The United States", *American Journal of Agricultural Economics,* Jan 21, 2010

Meeusen, W. and J. van den Broeck (1977), "Efficiency Estimation from Cobb-Douglas Production Functions with Composed Error", *International Economic Review,* 18, 435-444.

Miller Perry R., David E. Buschena, Clain A. Jones and Jeffrey A. Holmes (2008), "Transition from Intensive Tillage to No-Tillage and Organic Diversified Annual Cropping Systems", Published n *"Agron J"* 100:591-599 (2008) DOI: 10.2134/agronj2007.0190, available on the world wide web: http://agron.scijournals. org/cgi /content/abstract/100/3/591

Mohana Rao, L.K. and B. Prasada Rao (1988) "Studies in the Farm management in the Command Area of Nagarjuna Sagar Irrigation Project" *Directorate of Economics and Statistics, Department of Agriculture and Co-operation, Ministry of Agriculture, Govt. of India,* New Delhi

Mohana Rao. L.K (2011) Detailed discussion on the general economic development of India in the recent past, budget meet 2011 held at Dept. of Economics, Andhra University, Visakhapatnam

Nemecek, T; O. Hugnenin. Elie, D. Dubois and G. Gailord (2005) "Okobilanzierung von anbausystemen im schweizericschen Acker – und futterbau", *Schriftenreihe der* FAL, 58 FAL Reckenholz, Zurich

Niggli, U., A. Fliebach, P. Hepperly, J. hanson, D. Douds and R. Seidel (2009), "Low Greenhouse Gas Agriculture: Mitigation and Adoption Potential of Sustainable Farming System", *Food and Agriculture Organization, Review – 2,* pp.1-22.

Offermann F., Nieberg H. (a cura di) (2000) "Economic Performance of Organic Farms in Europe. Organic Farming in Europe", *Economics and Policy,* Volume 5. Hohenheim, Universität Hohenheim

Oude lansink A., Pietola K., Backman S., (2002) "Efficiency and productivity of conventional and organic farms in Finland 1994-1997", *European Review of Agricultural Economics,* Vol no. 29, 51-65.

Padel S and Uli Z (1994) "Economics of organic farming in Germany" in The economics of organic farming – An international perspective (ed) by Lampkin N.H and Padel S., *CAB International Publishers*

Parikh, A. and K.Shah(1994), " Measurement of Technical efficiency in the North West Frontier Provience of Pakistan", *Journal of Agricultural Economics,* 45, 132-138.

Pimentel David (2005) "Organic farming produces same corn and soybean yields as conventional farms, but consumes less energy and no pesticides", *Bioscience,* Vol 55:7, available on the world wide web: http://www.news. cornell.edu/stories/July05/organic.farm.vs.other.ssl.html

Pimentel,D., P. Hepperly, J. Hanson, D. Douds and R. Seidel (2005), "Environmental, Energetic and Economic Comparisons of Organic and Conventional Farming Systems", *Bioscience,* Vol.55 pp.573-582.

Pluke Richard,Amy Guptill (2004) "The Social, Ecological and Farming System Constraints on Organic Crop Protection in Puerto Rico" Paper presented at a symposium entitled *"IPM in Organic Systems"*, XXII International Congress of Entomology, Brisbane, Australia, 16 August 2004., available on the world wide web : http://www.organic-research.com/

Posnera Joshua L., Jon O. Baldock and Janet L. Hedtcke (2008), "Organic and Conventional Production Systems in the Wisconsin Integrated Cropping Systems Trials: I". Productivity 1990–2002, Published in "*Agron J*" 100:253-260 (2008), American Society of Agronomy, 677 S., Segoe Rd., Madison, WI 53711 USA, available on the world wide web: http://agron.scijournals.org/cgi /content/abstract/100/2/253

Prasad, R. (1999), Organic farming *vis-à-vis* modern agriculture *Curr. Sci.*, 1999, 77, 38–43.

Pretty, Jules and Ball Andrew (2001), Agricultural Influences on Carbon Emissions and Sequestration: A Review of Evidence and the emerging Trading Options, *Occasional Paper, Centre for Environment and Society and Department of Biological Sciences*, University of Essex, U.K.

Ramesh Chand and Raju S.S (2008) Instability in Indian Agriculture, *Discussion Paper: NPP01/2008*, National Centre for Agricultural Economics and Policy Research, Pusa, New Delhi.

Ramsamy, C. *et al.*, (2003) "Hybrid Rice in Tamil Nadu Evaluation of Farmers' Experience" *Economic and political Weekly*, June 21 2003, pp.2509-2512.

Rao,CH. H, (1965) "Agricultural Production Functions, Costs and Returns in India", Asia Publishers, Bombay

Rao,CH.H, (1975)"Technological Changes and Distribution of Gains in Indian Agriculture", Macmillan Publishers, New Delhi.

Rasmussen, P.E., K.W.T. Goulding, J. R. Brown, P. R. Grace, H.H. Janzen and M. Korschens (1998), "Long Term Agro-ecosystem Experiments: Assessing Agricultural Sustainability and Global Change", *Science*, Vol.282, pp.893-896.

Reganold, JP., Glover, JD., Andrews, PK ., and Hinman, HR, (2001), "Sustainability of three apple production systems" *Nature*. Vol. 410, no. 6831, pp. 926-930. 19 Apr 2001.

Regonald, J.P, L.F. Elliot and Y.L. Unger (1987), "Long-Term Effects of Organic and Conventional Farming on Soil Erosion", *Nature*, Vl.330, pp.370-372

Reicosky, D.C, W.D. Kemper, G. W. Langdale, C.L. Douglas and P.E. Rasmussen (1995), "Soil Organic Matter Changes Resulting From Tillage and Biomass Production," *Journal of Soil and Water Conservation*, Vol.50, No.3, pp.253-261.

Reifschneider, D. and R. Stevenson (1991), "Systematic Departures from the Frontier: A Framework for the Analysis of Firm Inefficiency", *International Economic Review,* 32, 715-723.

RobertM., J. Antoine and F. Nachtergaele (2001), *Carbon Sequestration in soils, Proposal for Land Management in Arid Areas of the Tropics,* AGLL, Food and Agriculture Organization of the United Nations, Rome, Italy.

Robertson, C.A, (1971) "Introduction to Agricultural Production Economics and Farm Management", Tata Macgrahil, New Delhi

Sala, O.E. and J.M. Paruelo (1997), "Ecosystem Services in Grasslands", in G. Daily (Ed) (1997), *Nature's Services: Societal Dependence on Natural Ecosystems,* Island Press, Washington, D.C., U.S.A.

Seiford, L.M and Thrall R.M (1990) "Recent Developments in DEA: The Mathematical Programming Approach to Frontier Analysis" *Journal of Econometrics,* 46, Pp: 7-38, North Holland

Sekar, C., C. Ramasamy and S.Senthilnathan (1994), "Size Productivity Relations in Paddy Farms of Tamil Naidu", *Agricultural Situation in India* 48: 859-863

Sharma, V.P and Sharma P (2002) Trade liberalization and Indian Diary Industry, CMA Monograph no.196, IIM, Ahmedabad, *Oxford and IBH Publishing Co.Pvt.Ltd,* New Delhi.

Shephard, R.W (1970) "Theory of cost and production functions", *Princeton University Press,* Princeton

Shirsagar KG (2008) "Impact of Organic Farming on Economics of Sugarcane Cultivation in Maharashtra", *Working Paper No.15,* Gokhale Institute of Politics and Economics, Pune

Siegrist, S., D. Staub, L. Pfiffner and P. Mader (1998) "Does Organic Agriculture Reduce Soil Erodibility? The Results of a Long-Term Field Study on Losses in Switzerland," *Agriculture, Ecosystems and Environment,* Vol.69, pp. 253-264.

Singh S. (2009) Organic produce supply chains in India: Organization and Governance, *CMA publication no.222,* IIM, Ahmedabad - 380015

Singh Y. V., B. V. Singh, S. Pabbi and P. K. Singh(2007) "Impact of Organic Farming on Yield and Quality of BASMATI Rice and Soil Properties" available on world wide web http://orgprints.org/9783

Singh.S.R, (1986) "Technological Parameters in Agricultural Production Function" Ashish Publishing Home, New Delhi.

Smith, K.A. (1999), "After Kyoto Protocol: Can Scientists Make a Useful Contribution?" *Soil Biol. Biochemistry,* Vol.15,pp.71-75.

Statistics and Emerging Trends 2010, The World of Organic Agriculture, IFAOM, Bonn and FiBL, Frick

Stolze Matthias, Nicolas Lampkin (2009) "Policy for organic farming: Rationale and concepts" Published by "*ELSEVIER*". 0306-9192/$ - see front matter 2009 doi:10.1016/j.foodpol.2009.03.005 and also available on the world wide web: http://www.sciencedirect.com/science/article/B6VCB-4W2M6V2- ff57298e

Swezey Sean L.,Polly Goldman,Janet Bryer,Diego Nieto (2004), "Comparison Between Organic, Conventional, and IPM Cotton in the Northern San Joaquin Valley, California", Paper presented at a symposium entitled *"IPM in Organic Systems"*, XXII International Congress of Entomology, Brisbane, Australia, 16 August 2004 available on the world wide web: http://www.organic-research.com/

The Organic Standard and The Agricultural and Processed Food Products Export Development Authority (APEDA) 2010.

Tilman, D. (1998), "The Greening of the Green Revolution", *Nature*, Vol.396, pp.211-212.

Tzouvelekas V., Pantzios C.J., Fotopoulos C. (2002b), "Empirical Evidence of Technical Efficiency Levels in Greek Organic and Conventional Farms", *Agricultural Economics Review* 3: 49-60.

Tzouvelekas V., Pantzios C.J., Fotopoulos C., (2001a) "Technical efficiency of alternative farming systems: the case of Greek organic and conventional olive-growing farms", *Food Policy 26*, 549-569.

Tzouvelekas Vangelis, Christos J. Pantzios, and Christos Fotopoulos (2001), "Economic Efficiency in Organic Farming: Evidence from Cotton Farms in Viotia, Greece" *Journal of Agricultural & Applied Economics.*, Volume 33, April, 2001, Issue: 1, Pp: 35-48, available on the world wide web: http://ideas.repec.org/a/jaa/jagape/v33y2001i1p35-48.html#abstract

UNDP (1992), Benefits of Diversity: An Incentive towards Sustainable Agriculture, *United Nations Development Programme*, New York

Wood Richard, Manfred Lenzen, Christopher Dey, Sven Lundie (2005) "A comparative study of some environmental impacts of conventional and organic farming in Australia" Published in "*ELSEVIER*". and also available on the world wide web: http:// doi:10.1016/j.agsy.2005.09.007

Worthington M.K. (1980) Problems of Modern Farming, *Food Policy*, August issue.

Wynen E (1994) "Economics of organic farming in Australia" in The economics of organic farming – An international perspective (ed) by Lampkin N.H and Padel S., *CAB International Publishers.*

Wyss E.,H. Luka,L. Pfiffner,C. Schlatter,G. Uehlinger,C. Daniel (2004), "Approaches to Pest Management in Organic Agriculture: A Case study in European Apple Orchards" Paper presented at a symposium entitled

www.ingramcontent.com/pod-product-compliance
Ingram Content Group UK Ltd.
Pitfield, Milton Keynes, MK11 3LW, UK
UKHW040241300726
14061UKWH00001BD/86

9 789351 308348

"IPM in Organic Systems", XXII International Congress of Entomology, Brisbane, Australia, 16 August 2004, available on the world wide web: http://www.organic-research.com/

Yadav C.P.S., Harimohan Gupta, Dr. R. S. Sharma, *Organic Farming and Food Security: A Model for India*, Organic Farming Association of India, 2010.